Computationa

Endorsed by the *International Society for Computational Biology*, the *Computational Biology* series publishes the very latest, high-quality research devoted to specific issues in computer-assisted analysis of biological data. The main emphasis is on current scientific developments and innovative techniques in computational biology (bioinformatics), bringing to light methods from mathematics, statistics and computer science that directly address biological problems currently under investigation.

The series offers publications that present the state-of-the-art regarding the problems in question; show computational biology/bioinformatics methods at work; and finally discuss anticipated demands regarding developments in future methodology. Titles can range from focused monographs, to undergraduate and graduate textbooks, and professional text/reference works.

Fahad Saeed · Muhammad Haseeb

High-Performance Algorithms for Mass Spectrometry-Based Omics

Springer

Fahad Saeed
Knight Foundation School of Computing
and Information Sciences
Florida International University
Miami, FL, USA

Muhammad Haseeb
Knight Foundation School of Computing
and Information Sciences
Florida International University
Miami, FL, USA

ISSN 1568-2684 ISSN 2662-2432 (electronic)
Computational Biology
ISBN 978-3-031-01962-3 ISBN 978-3-031-01960-9 (eBook)
https://doi.org/10.1007/978-3-031-01960-9

This Springer imprint is published by the registered company Springer Nature Switzerland AG
The registered company address is: Gewerbestrasse 11, 6330 Cham, Switzerland

Fahad Saeed dedicates this book to Saba, Haadi, and Emaan.

Preface

To date, the processing of high-throughput Mass Spectrometry (MS) data is primarily accomplished using serial algorithms. Developing new methods to process MS data is an active area of research [1], but there is no single strategy that focuses on *scalability* of MS-based methods [2]. MS is a diverse and versatile technology for high-throughput functional characterization of proteins, small molecules, and metabolites in complex biological mixtures. In the recent years, the technology has rapidly evolved and is now capable of generating increasingly large (multiple terabytes per experiment) [1] and complex (multiple species/microbiome/high-dimensional) data sets [3]. This rapid advances in MS instrumentation must be matched by equally fast and rapid evolution of scalable methods developed for the analysis of these complex data sets. Ideally, the new methods should leverage the rich heterogeneous computational resources available in a ubiquitous fashion in the form of multicore, manycore, CPU-GPU, CPU-FPGA, and IntelPhi architectures. The absence of these high-performance computing algorithms now hinders scientific advancements in MS research [2].

In systems, biology setting workflows (or pipelines) are frequently used which are a sequence of loosely connected computational tasks. Database-search workflows are the most commonly used data processing pipelines which require matching a high-dimensional noisy MS data (called spectra) to a database of protein sequences. The entire workflow is executed using as a script-like structure that executes different algorithms which is then run on a dedicated workstation. The data volume can easily reach terabyte level depending on the experiment and search parameters for these workflows. The currently used state-of-the-art serial and parallel methods are data (and communication cost) oblivious which may not give the best possible performance for these database-search workflows. Currently used state-of-the-art serial algorithms results in unusually long processing times.

Development of parallel computing techniques to deal with this deluge of data has also been limited and has mostly adopted the batch mode (or embarrassingly parallel computing) form of processing. As one might imagine there have been some efforts in developing HPC algorithms that can be used for speeding up the processing. However, most of these efforts have been limited to parallelization of specific algorithms or workflows without the ability to generalize the end-to-end performance for other existing or new algorithms. In order to take the field of computational proteomics forward and set it afoot with more mature fields such as genomics more concerted HPC efforts. This has resulted in limited speedups (e.g. 30x speedup for 200 cores) and consequently limited usage by proteomics practitioners who might not see the advantage of trivial reduction in processing times.

Our own preliminary study suggests that such workflows when used with data divided among compute nodes in an oblivious manner lead to unbalanced workload and results in hours to weeks of computation with the state-of-the-art software [4]. Therefore, concerted efforts are needed for the development of high-performance computing (HPC) framework for working with large mass spectrometry data sets while benefiting advancements in areas of science including proteomics, proteogenomics, meta-proteomics, and microbiomes. These frameworks must be able to leverage the vast HPC heterogeneous architectures that are ubiquitous in the form of desktops, laptops, clusters, and supercomputers.

The scientific premise of this project is that progress can be gained in developing scalable MS-based omics data analysis tools for non-model organism proteomics/meta-proteomics/proteogenomic will require: (1) Improved data-partitioning strategies allowing minimization of data communication between different levels of memory hierarchy and processing units; (2) Improved parallel algorithms on distributed-memory architectures to address the scalability limitations due to excessive communication costs; (3) Improved parallel algorithms for CPU-GPU/CPU-FPGA architecture to exploit heterogeneity of modern HPC machines; (4) Integration of these parallel algorithms to XSEDE Supercomputers Gateways will make our methods available for large-scale omics system biology studies.

To address these challenges, we will formulate and develop MS-specific parallel computing abstraction that incurs minimal I/O times on the multitude of heterogeneous architectures. Second, to address the diversity of the mass spectrometry user community, we will formulate HPC frameworks that supports scaling down analysis (i.e. working with large data files on relatively inexpensive hardware such as CPU-GPU without fully loading them into memory), as well as scaling up (i.e. ability to execute the workflows on large XSEDE supercomputers and cloud computing infrastructure). By ensuring the compatibility of mass spectrometry-specific data standards, supporting the number of heterogeneous architectures for efficient processing and generating optimized code bases in open-source format will enable the development of scalable analytical methods. Therefore, our framework aims to democratize access to HPC infrastructure for a broader community of life and systems biology scientists, and create a blueprint for a new paradigm for HPC computing for large MS data sets—making HPC the fourth pillar of scientific investigations.

Miami, FL, USA
January 2022

Fahad Saeed

References

1. Kong AT, Leprevost FV, Avtonomov DM, Mellacheruvu D, Nesvizhskii AI (2017) Msfragger: ultrafast and comprehensive peptide identification in mass spectrometry–based proteomics. Nat Methods 14(5):513
2. Haseeb M, Afzali F, Saeed F (2019) Lbe: A computational load balancing algorithm for speeding up parallel peptide search in mass-spectrometry based proteomics. In: IEEE International parallel and distributed processing symposium workshops (IPDPSW), IEEE, 2019, pp 191–198
3. Tariq MU, Saeed F (2021) Specollate: Deep cross-modal similarity network for mass spectrometry data based peptide deductions. PloS one 16(10), e0259349
4. Awan MG, Saeed F (2016) Ms-reduce: an ultrafast technique for reduction of big mass spectrometry data for high-throughput processing. Bioinformatics 32(10):1518–1526

Acknowledgements

The research presented was supported by the NIGMS of the National Institutes of Health (NIH) under award number: R01GM134384. The authors were further supported by the NSF under award number: NSF CAREER OAC-1925960. The content is solely the responsibility of the authors and does not necessarily represent the official views of the NIH and/or the NSF. This work used the National Science Foundation (NSF) XSEDE supercomputers through allocations TG-CCR150017 and TG-ASC200004. The authors would also like to acknowledge hardware donation by Intel Altera (DE10-PRO-SX FPGA) and NVIDIA (TITAN Xp GPU).

Contents

Chapter 1
Need for High-Performance Computing for MS-Based Omics Data Analysis

Fahad Saeed and Muhammad Haseeb

For the past 30 years, significant efforts were invested for the development of designing and implementing more efficient scoring functions which included highly successful search engines [1–5]. Similar to other domains, numerical algorithms were developed for Mass Spectrometry (MS)-based peptide deduction and are designed and implemented by assuming *number of arithmetic operations* as the sole metric for efficiency. In the last decade, the technological trend of Moore's law has kept making the arithmetic operations faster. As a result, the bottleneck for many algorithms have shifted from computational arithmetic operations efficiency to communication, i.e. communication costs of either moving the data between different levels of memory hierarchy (e.g. RAM, cache) or between different distributed-memory processors connected via a network. There are numerous [6–8] studies that have shown this trend that the cost of moving data exceeds the costs of doing the arithmetic operations [6]. This gap is, and will continue to grow exponentially over time with the introduction of multicore, manycore, and GPU architectures [8–10].

To date, processing of high-throughput MS data is accomplished mostly using serial algorithms. Recent trends in systems biology (e.g. meta-proteomics, non-model proteomics, microbiomes) MS-based experiments point towards the need for larger, better, and faster computational tools. This trend is fed by the MS technology that has rapidly evolved and is now capable of generating increasingly large and complex (multiple species/microbiome/high-dimensional) data sets [11]; leading to high-impact proteomics, meta-proteomics and microbiomes studies directly related to human disease and health. This rapid advances in MS instrumentation must be matched by equally rapid evolution of scalable methods developed for the analysis of these complex data sets. Developing new methods to process MS data is an active area of research [5] but the focus of this research, till date, has been towards improving the efficiency of arithmetic operations. Exclusion of communication costs as a metric, of otherwise highly successful methods, is becoming a severe bottleneck for

F. Saeed and M. Haseeb, *High-Performance Algorithms for Mass Spectrometry-Based Omics*, Computational Biology 34,
https://doi.org/10.1007/978-3-031-01960-9_1

processing of MS data, and now hinders scientific advancements for microbiome, and (meta) proteomics/microbiome research [12] workflows are the most commonly used data processing pipelines that match the high-dimensional noisy MS data (called spectra) to a database of protein sequences. These MS data sets are then processed using databases which may be several times larger than the original proteome (or a combination of multiple proteomes in case of meta-proteomics studies) depending on the search parameters. The data volume can easily reach terabyte level depending on the experiment and search parameters for these workflows.

Comparison across literature indicates decreased scalability of the serial algorithms. The scalability issues for these serial methods are solved using various filtering mechanisms but this has shown increased misidentification of peptides and/or inconsistencies between various search engines [13, 14]. It is not uncommon to have a terabyte scale database against which millions of raw experimental spectra need to be processed for model organisms (e.g. Rattus) and a few dynamic modification as search parameters. Non-model organism is considered the next frontier to accelerate insight into chronic disease in humans [15] requires even larger search spaces which would lead to intractable runtimes for both serial algorithms and traditional HPC methods for MS-based omics.

Increasing size of the spectra and theoretical database search space has led to the development of high-performance computing strategies [16–21] to speed up these search engines. However, similar to serial numerical algorithms, the objective of these HPC methods has been to speed up the arithmetic scoring part of the search engines with little to no efforts to minimize the communication costs. However, within-study direct comparisons of serial versus high-performance computing (HPC) proteomics algorithms have revealed modest gains [22]. Cloud computing-based processing, alone, cannot fill the gap of scalable tools since a non-efficient tool on desktop or cluster is non-efficient on cloud and scaling the cloud resources are costly prohibitive. All these indicate that further formal design and evaluation are warranted for scalable infrastructure for MS-based omics database workflows. High-performance computing algorithms that can leverage multicore, manycore, distributed-memory, and CPU-GPU architectures can make the existing pipelines much more scalable while maintaining a high confidence in peptide identifications. More modern techniques based on deep learning models [23, 24] are known to have scalability problems with increasing size of the training sets and will incur similar bottlenecks when incorporated into regular MS-based omics workflows.

Therefore, there is an urgent need for scalable solutions of more confident peptide identifications without which the integrity, and the confidence in large-scale systems biology studies is not possible, especially for meta-proteomics, proteogenomics, and MS-based microbiome or non-model organisms' studies having a direct impact on personalized nutrition, microbiome research, and cancer therapeutics.

In order to fill this gap, we collectively as a scientific community have to design parallelization techniques, and approaches for high-performance MS omics data analysis. However, in contrast to existing methods, these HPC algorithms must be designed by considering both computational, and communication costs as metrics for efficiency. One parallel algorithm[1] with provable efficiencies can be designed for distributed-memory architectures, it can be extended for multicore, manycore, and Graphics Processing Units (GPU's), and cloud-platforms. To maximize the impact, effective research works are required to solve the following facets: (1) Improved data-partitioning strategies allowing minimization of data communication between different levels of memory hierarchy, and processing units; (2) Improved parallel algorithms on distributed-memory architectures to address the scalability limitations due to excessive communication costs, and (3) Integration of these parallel algorithms to existing workflows using XSEDE Supercomputers, and Amazon Cloud-Computing infrastructure.

Additional capabilities expected from highly scalable HPC methods include new-found abilities to process: (1) large number of potential dynamic PTM's search parameters, (2) Limiting filtering that may lead to omics dark-data, and (3) search against multiple organisms from different taxonomic families for meta-proteomics and microbiome studies. Integration of these HPC methods to existing workflows will enable systems biologist to investigate complex microbiome communities and identify the taxonomic families which is so intimately associated with human health and disease. This novel, substantially different, and scalable computational approach to study complex proteomes (and microbiome) at the proteomics level is expected to allow us to overcome the current scalability limitations of existing workflows. This new class of HPC algorithms, we believe, will open these new horizons for precision nutrition studies, interaction insights for complex microbiome communities and its effects on human gut, and overall mental and physical health.

The proposed HPC research for MS-based omics data analysis is innovative, in our opinion, because it represents a substantive departure from the status quo. To our knowledge, communication-avoiding parallel algorithms that consider both computation and communication costs to improve the efficiency of these workflows has never been achieved in the context of MS-based omics data analysis. Although rarely employed to date, the incorporation of novel and exciting parallel algorithmic design will make the power of supercomputing, ubiquitous manycore, and GPU-based architectures accessible to large scores of systems biology scientists. Such an accessible HPC framework will serve as the proof-of-concept infrastructure which will *enable bold questions*, and large system biology studies not hindered or hampered by the limiting factors associated with prohibitively long running times and/or non-accessible workflows.

Our hope is that HPC will help in understanding, and studying microbiome MS-based omics in the same way it has made a remarkable difference in our understanding of the cosmos, genomics, and molecular dynamics.

---

[1] Parallel algorithm and high-performance computing (HPC) method will be used interchangeably in this book.

References

1. Eng JK, Fischer B, Grossmann J, MacCoss MJ (2008) A fast sequest cross correlation algorithm. J Proteome Res 7(10):4598–4602
2. Diament BJ, Noble WS (2011) Faster sequest searching for peptide identification from tandem mass spectra. J Proteome Res 10(9):3871–3879
3. Eng JK, McCormack AL, Yates JR (1994) An approach to correlate tandem mass spectral data of peptides with amino acid sequences in a protein database. J Am Soc Mass Spectrom 5(11):976–989
4. McIlwain S, Tamura K, Kertesz-Farkas A, Grant CE, Diament B, Frewen B, Howbert JJ, Hoopmann MR, Kall L, Eng JK et al (2014) Crux: rapid open source protein tandem mass spectrometry analysis. J Proteome Res 13(10):4488–4491
5. Kong AT, Leprevost FV, Avtonomov DM, Mellacheruvu D, Nesvizhskii AI (2017) Msfragger: ultrafast and comprehensive peptide identification in mass spectrometry-based proteomics. Nat Methods 14(5):513
6. Ballard G, Demmel J, Holtz O, Schwartz O (2011) Minimizing communication in numerical linear algebra. SIAM J Matrix Anal Appl 32(3):866–901
7. Council NR et al (2005) Getting up to speed: the future of supercomputing. National Academies Press
8. Ballard G, Carson E, Demmel J, Hoemmen M, Knight N, Schwartz O (2014) Communication lower bounds and optimal algorithms for numerical linear algebra. Acta Numerica 23:1
9. Demmel J, Eliahu D, Fox A, Kamil S, Lipshitz B, Schwartz O, Spillinger O (2013) Communication-optimal parallel recursive rectangular matrix multiplication. In: 2013 IEEE 27th international symposium on parallel and distributed processing. IEEE, pp 261–272
10. Solomonik E, Bhatele A, Demmel J (2011) Improving communication performance in dense linear algebra via topology aware collectives. In: SC'11: Proceedings of 2011 international conference for high performance computing, networking, storage and analysis. IEEE, pp 1–11
11. Saito MA, Bertrand EM, Duffy ME, Gaylord DA, Held NA, Hervey WJ, Hettich RL, Jagtap PD, Janech MG, Kinkade DB, Leary DH, McIlvin MR, Moore EK, Morris RM, Neely BA, Nunn BL, Saunders JK, Shepherd AI, Symmonds NI, Walsh DA (2019) Progress and challenges in ocean metaproteomics and proposed best practices for data sharing. J Proteome Res 18(4):1461–1476, pMID: 30702898. http://dx.doi.org/10.1021/acs.jproteome.8b00761
12. Yates III JR (2019) Proteomics of communities: metaproteomics
13. Griss J, Perez-Riverol Y, Lewis S, Tabb DL, Dianes JA, del Toro N, Rurik M, Walzer M, Kohlbacher O, Hermjakob H et al (2016) Recognizing millions of consistently unidentified spectra across hundreds of shotgun proteomics datasets. Nat Methods 13(8):651
14. Tran NH, Zhang X, Xin L, Shan B, Li M (2017) De novo peptide sequencing by deep learning. Proc Natl Acad Sci 114(31):8247–8252
15. Heck M, Neely BA (2020) Proteomics in non-model organisms: a new analytical frontier. J Proteome Res
16. Kulkarni G, Kalyanaraman A, Cannon WR, Baxter D (2009) A scalable parallel approach for peptide identification from large-scale mass spectrometry data. In: 2009 international conference on parallel processing workshops. IEEE, pp 423–430
17. Li C, Li K, Li K, Lin F (2019) Mctandem: an efficient tool for large-scale peptide identification on many integrated core (mic) architecture. BMC Bioinform 20(1):397
18. Sun J, Chen B, Wu F-X (2014) An improved peptide-spectral matching algorithm through distributed search over multiple cores and multiple cpus. Proteome Sci 12(1):18
19. Duncan DT, Craig R, Link AJ (2005) Parallel tandem: a program for parallel processing of tandem mass spectra using pvm or mpi and x! tandem. J Proteome Res 4(5):1842–1847
20. Bjornson RD, Carriero NJ, Colangelo C, Shifman M, Cheung K-H, Miller PL, Williams K (2008) X!! tandem, an improved method for running x! tandem in parallel on collections of commodity computers. J Proteome Res 7(1):293–299

21. Li C, Li K, Chen T, Zhu Y, He Q (2019) Sw-tandem: a highly efficient tool for large-scale peptide sequencing with parallel spectrum dot product on sunway taihulight. Bioinformatics (Oxford, England) 35(19):3861–3863
22. Saeed F, Haseeb M, Iyengar S (2020) Communication lower-bounds for distributed-memory computations for mass spectrometry based omics data. arXiv:2009.14123
23. Qiao R, Tran NH, Li M, Xin L, Shan B, Ghodsi A (2019) Deepnovov2: better de novo peptide sequencing with deep learning. arXiv:1904.08514
24. Chi H, Liu C, Yang H, Zeng W-F, Wu L, Zhou W-J, Niu X-N, Ding Y-H, Zhang Y, Wang R-M et al (2018) Open-pfind enables precise, comprehensive and rapid peptide identification in shotgun proteomics. bioRxiv, 285395

Chapter 2
Introduction to Mass Spectrometry Data

Fahad Saeed and Muhammad Haseeb

Mass spectrometry (MS) is used to elucidate the chemical structures of peptide molecules and has numerous systems biology applications [1–12]. Mass spectrometry is also used in metabolomics, glycomics, lipidomics, and clinical applications [5, 6, 13–16]. In the first stage for mass spectrometry-based proteomics, unknown proteins are isolated and digested using enzymes (such as trypsin) to cut these proteins into smaller pieces called peptides. These peptides are unknown and the masses of these peptides are separated using high-performance liquid chromatography (HPLC) and thereafter fragmented using two stage mass spectrometry (MS/MS or MS2) to produce millions of these mass spectra known as MS2. The new mass spectrometry machines such as Thermo Orbitrap Fusion can produce millions of MS2 spectra within hours [11, 29]. Protein molecules are combinatorial chains of 20 basic amino acids. And study of proteomics mostly deals with sequencing of these chains.

We discuss few of the MS-based omics strategies, their data generation, and methods with which most of the analysis is completed.

2.1 Proteomics

2.1.1 Mass Spectrometry-Based Proteomics

Two widely used approaches to analyze proteins using mass spectrometry from complex biology samples are Top-Down approach and Bottom-up approach. Both techniques allow one to get data that is specific to the proteins but the data that is acquired has to be analyzed in different formats and with different assumptions. In the Top-Down approach the purified samples of proteins are fed into the mass spectrometer

Parts of this chapter may have appeared in [24].

F. Saeed and M. Haseeb, *High-Performance Algorithms for Mass Spectrometry-Based Omics*, Computational Biology 34,
https://doi.org/10.1007/978-3-031-01960-9_2

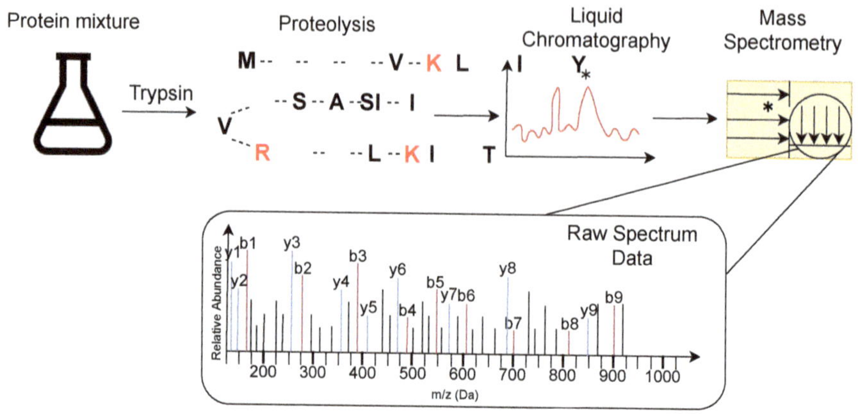

Fig. 2.1 Process of generating MS/MS spectra from a protein mixture using mass spectrometry analysis. Protein in the mixture is broken into peptides using the enzyme called trypsin which breaks the protein strings at K and R bases generating peptides of varying sizes. This peptide mixture is then refined and peptides are moved through mass spectrometer which generates an MS/MS spectrum for each peptide of different mass

where they are ionized, fragmented, and ultimately "read" by the mass spectrometer. The resulting data is tandem mass spectrum which is then analyzed using plethora of computational methods for identification of proteins, and the variants that might occur. The second most commonly used method called Bottom-Up approach or Shotgun approach works as follows. Purified protein mixture is proteolyzed into a peptides using various enzymes (e.g. Trypsin). This peptide mixture is then sent to the mass spectrometer which using automated liquid chromatography mass spectrometry LC-MS/MS pipeline is used for fragmentation, and measuring the mass-to-charge ratio of each of the fragments. For the measurements using mass spectrometer, in the first stage the proteins are ionized, and MS1 spectra is acquired. Thereafter, the peptide ions are selected based on the m/z and fragmented which are then depicted as MS2 spectra. A simplified version of Mass Spectrometry is shown in Fig. 2.1. The resulting MS2 spectra are then processed using computational tools in targeted or untargeted manner to identify specific or all peptides in a given sample.

The two data acquisition methods in the shotgun LC-MS/MS pipeline include:

2.1.1.1 Data Dependent Acquisition (DDA)

The peptide eluting from the liquid chromatography (LC) at any given point in time is first detected at the first level of mass spectrum (MS1) in Data Dependent Acquisition (DDA). After this first step, k ($\sim$ 10–20) most abundant peaks (also known as peptide precursor ions) are separated using m/z which are further selected for fragmentation. Since the spectra that are acquired are very much dependent on the data that is available to LC at that point in time this technique is known as "data dependent". This data acquisition technique is advantageous because it allows separation of peptide ions in the first which greatly reduces the probability of so-called convoluted spectra and where each fragment-ion has a direct correlation to the peptide precursor-ion mass. Although the data that is acquired is cleaner that can be used for further interpretation; the data itself is less reproducible because peptide ions that are eluted have variable intensities across experiments. This method of data acquisition also suffers from low coverage of the proteins in the sample because a significant number of peptides ions are discarded in the first stage. Finally, the method may also suffer from low data-production rate when greater number of peptide ions are chosen for fragmentation in MS1 stage of the acquisition.

2.1.1.2 Data Independent Acquisition (DIA)

Data Independent Acquisition (DIA) operates by including all the peptides ion in the MS1 spectra that are within a constrained m/z ratio window that are then chosen for MS2 fragmentation [17]. If all the ions that are entering the mass spectrometer at a given point in time are chosen for fragmentation than such acquisition is called broadband DIA. Commonly used DIA method includes performing a MS^E fragmentation such that the peptide is first subjected to low-energy Collision-induced dissociation (CID) that is then followed by high energy fragmentation (e.g. Higher energy collisional dissociation (HCD) [17].

Another variant of DIA method which can produce significant amount of data is Sequential Window Acquisition of all Theoretical Mass Spectra (SWATH-MS) [18] which works by isolating swaths of specific m/z ranges in the MS1 spectra. The fragments of all peptides that are in each window for each point in the chromatogram. The ion chromatograms for fragment-ions in MS2 spectra and their precursor ions are extracted, called extracted ion chromatograms (XICs), which allow a drastic improvement of signal-to-noise ratio in the multiplexed MS2 data. The major advantage of SWATH-MS is that the windows can be fragmented in parallel which allow massively parallel high-throughput data. This enables accurate reproducibility, and significantly better coverage of the peptides in the sample [19].

This allows a complete digital profile of the sample for each time point in the chromatogram, co-eluting window for MS2 spectra for each MS1 spectrum. The drawback of this kind of approach is that MS2 data is highly convoluted because MS2 spectrum can now contain fragment-ions from multiple peptide precursors. This in turn leads to loss of direct one-to-one mapping of fragment-ion and precursor-ion [17, 18]. Therefore, to analyze this complex data sets several computational techniques have been proposed [20]. Many of these techniques including DIA-UMPIRE [21], DeMux [20], Group-DIA [22], MSPLIT-DIA [23] work by first detecting precursor and fragment-ion peaks from the multiplexed DIA spectra which leads to pseudo-tandem DDA-like MS/MS spectra. This pseudo-spectra is then used and searched against a sequence database or spectral library for identification. Other tools like PIQED [24] improve upon the previous methods by incorporating techniques to identify post-translational modifications (PTM) in DIA MS2 data. More recently, deep learning techniques for denovo identification of peptide sequences have been proposed in the form of methods such as DeepNovo-DIA [25] to identify peptides from these complex spectra. Open-SWATH [26] also allows targeted analysis of the DIA data. The reader is referred to this comprehensive survey [27] on strategies, methods, and software solutions for DIA data.

2.1.2 *MS/MS Data Pre-processing*

Any DDA MS/MS data obtained from mass spectrometer contains a combination of real signal and noise irrespective of the resolution of the mass spectrometer [28]. Our earlier studies have found that more than 90% of the spectra consist of noise [29]. Due to the process using which MS data is generated, similar peptide ions can get fragmented multiple times which can generate redundant data. This data can be considered analogous to coverage in next-generation genome sequencing in genomics and can be used for many statistical analysis, and clustering. Therefore, removing noisy peaks, low-quality spectra, reconstruction of missing peaks, and merging of redundant MS/MS spectra which can be further used for analysis [29–34], and may result in quicker analysis of data [31]. Some interesting tools include pParse [35] which using machine learning to classify co-eluting peptides, MS-REDUCE [29] that uses quantization, and random sampling for noise reduction, MS-Clustering [34], Pep-Miner [36], CAMS-RS [37], MaRaCluster [38] that incorporate different metrics to cluster raw MS data which can be used for further processing. A more comprehensive survey for pre-processing MS/MS data is available in [28].

2.1.3 *Peptide Identification*

Deduction of peptides from MS data is one of the fundamental operations that are executed for DDA tandem MS/MS spectra. There are three main techniques [39] that are used for this deduction: (1) Donovo methods; (2) spectral library search methods; (3) database search methods.

2.1.3.1 De Novo

In an event there is no reference library, or if non-model organism has been used for experiments; De novo peptide deduction method is utilized [40]. The peaks in the spectra are used for labeling amino acid masses, charge, and fragmentation that can lead to deduction of peptides. However, most of the De Novo methods result in low accuracy, or confidence which result in low identification rate for highly challenging PTM spectra, fragmentation errors, and mutations in the peptide identification processes. PepNovo [40], pNovo [41], Open-pNovo [42], DeepNovo [43], PEAKS [44],

Lutefisk [45], Spectral Networks [46], AuDeNs [47], MSNovo [48], SeqMS [49], PFIA [50], and NovoHMM [51] are some of the Denovo methods that are currently available. Most of these tools construct graphs where the vertices are the peaks, and the optimal path between them is constructed using dynamic programming based on fragmentation model. To date, Denovo methods have not found widespread usage due to accuracy and confidence lmitations.

2.1.3.2 Spectral Library Search

When a library of annotated MS data is available, the experimentally acquired MS data can be compared against such a library using statistical techniques, and signal processing methods. The advantage of spectral library search is accurate peptide identification, and quantification of PTM's [52]. The disadvantages include small size of spectral libraries, the need for annotation (some can be done computationally but needs a human to supervise), and generally the need to perform similar experimental conditions for the spectra to confidently match with the library. ANN-SoLo [53], pMatch [54], SpectraST [52], Pepitome [55], NIST MSPepSearch [56], X! Hunter [57], ProMEX [58], HMMatch [59], MSDash [60], QuickMod [61], OMSSA [62], and MzMod [63] are some of the published work in this area where researchers have shown high rate of accuracy for a variety of conditions, and PTMs.

2.1.3.3 Database Search

The most accurate, and commonly used method suitable for peptide identification is database search methods which operate by comparing experimental MS/MS spectra against predicted spectra (from model organisms databases) [64]. The theoretical spectral libraries are first produced by digesting in-silico proteome sequence databases into short peptide sequences, and then predicting the MS/MS spectra using the m/z ratio, and the charge of the amino acids. Post-translational modifications that are part of the search-parameters are also used for the prediction of the individual peaks in the (modified) spectra. Recently, few machine learning techniques [65–68] have also been incorporated in these database search tools that improve on the prediction of the peaks for different fragmentations, and have been shown to improve the accuracy of the database search tools due to superior matching of the spectra against these theoretical spectra. Most commonly used search tools include SEQUEST [69], SpecOMS [70], MSFragger [5], Open-pFind [71], Andromeda [72], X! Tandem [73], Crux [74], Tide [75], Comet [76], TagGraph [77], PEAKS-DB [78], and JUMP [79].

The theoretical protein database can expand in the search space that is required for matching the spectra, and many of these tools employ various filtration techniques that can be used for efficient computations which are usually followed by peptide-to-spectrum matches. There is also significant work in developing scoring techniques that are fast and can be used for massive data sets which include shared-peak counting [5, 62, 70, 80–82], sequence tagging [71, 77, 78, 83–88], and peptide precursor mass

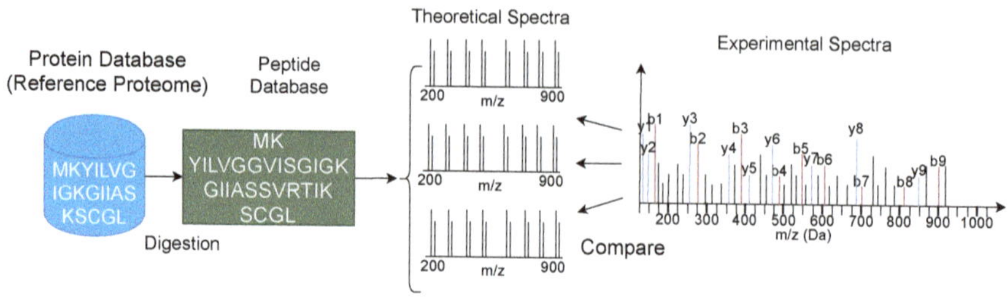

Fig. 2.2 A generic proteomics flow. In-silico digestion of the protein database is performed to generate peptides. These peptides are then converted to the theoretical spectra and compared against the experimental spectra

[76, 89–91] based methods. These methods use various features of the spectra data such as precursor charge, enzyme specificity, and other features in the MS data to filter the search space for efficient computations. The reduced search space is then generally used for scoring the peptide for a given spectra as demonstrated by various tools [62, 69, 73–76, 92–95].

Finally, the obtained peptide-to-spectrum match, spectrum to spectrum match and the de novo constructed tag scores are re-analyzed to assign confidence to the computed scores. In general, confidence scores are assigned using probabilistic classification models, a score also called False Discovery Rate (FDR), that computes the probably of a “true” match from a random chance match. In order to calculate FDR, tools such as Percolator [96–98], Protein Prophet [99], and iProphet [100] generate a decoy database which is then used to search MS spectra against decoy proteins which are eventually scored to assess the confidence in the matches. There are few other methods (TagGraph [77], Open-pFind [71]) that have been proposed which employ a decoy-free model and incorporate a probabilistic-distribution based models that have learned features from MS/MS data. A generic workflow of protein database search is shown in Fig. 2.2.

2.2 Proteogenomics

Proteogenomics studies are performed to discover novel peptides by searching the spectra against any of the conceivable peptide from six-frame, splice-graph or other customizable databases. The identified peptides are either intergenic (i.e. mapping to areas in DNA located between annotated genes) or intragenic (mapping to DNA located with in an annotated gene). The gene that has been annotated can be used to identify peptides that can map to exon, intron, alternative exon-exon junction, exon-intron junction, 5’ UTR, 3’ UTR sites [101]. A detailed proteogenomic workflow is shown in Fig. 2.3 which is depicting a 6-frame translation of the database search used for searching intragenic peptides performed using the theoretical protein database obtained from six-frame translation. Few limitations of this workflow

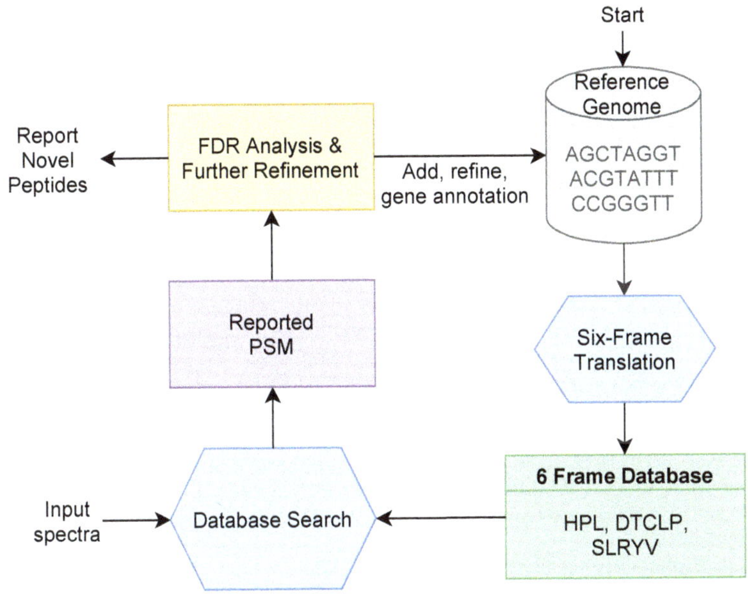

Fig. 2.3 Detailed flow for 6-frame database search. The process starts by generating a custom protein database by translating all the six frames. Experimental spectra are searched against this database to find novel peptides. The output from these tools is further refined using different tools. At the end, the novel discoveries are reported and can be used to add annotation to the reference genome or refine gene models

include using huge database size of mostly non-existing proteins which can lead to inflated number of peptide matches, which can also lead to inability to identify exon-exon junction peptides. There are some techniques that have been designed to increase the accuracy of these algorithms for matching such as creating customized database for proteogenomic analysis, and using filtration techniques to predict coding genes in the translated database. RNA-sequencing data can also be used to generate a more compact customized protein database.

Increasing number of proteogenomic tools [102–106] have been proposed till date which allows searching spectra against a custom or six-frame translated database, visualization, and automated annotation. First step includes obtaining a database against which experimental spectra can be searched and such a database can be constructed using either translating the entire genome (with all possible amino acid sequences) or using smaller databases such as pseudogene db, lncRNA db, splice db, UTR db, and sORF. Next, the experimental spectra are matched with this generated database to find novel peptides or proteins. Various filtering mechanism have been formulated to reduce the number of peptides that are searched in this large theoretical database to increase the confidence of the matches. Once, the identified peptides are selected, they are further curated and screened to remove any false matches that might have passed the FDR filter. To reduce the number of false positive matches against a very large- and inflated. As the search space for proteogenomic workflows is huge (hundreds of times larger than the reference proteome), the number of false positives tend to be much higher than the specified limit. Therefore, certain guidelines need to be followed when reporting proteogenomic matches as novel peptides [101].

References

1. Musbacher N, Schreiber TB, Daub H (2010) Glycoprotein capture and quantitative phosphoproteomics indicate coordinated regulation of cell migration upon lysophosphatidic acid stimulation. Mol Cell Proteomics 9(11):2337–2353. arXiv:www.mcponline.org/content/9/11/2337full.pdf+html, https://doi.org/10.1074/mcp.M110.000737http://www.mcponline.org/content/9/11/2337.abstract
2. Solit DB, Mellinghoff IK (2010) Tracing cancer networks with phosphoproteomics. Nat Biotech 28(10):1028–1029. https://doi.org/10.1038/nbt1010-1028
3. Gruhler A, Olsen V, Mohammed S, Mortensen P, Faergeman J, Mann M, Jensen N (2005) Quantitative phosphoproteomics applied to the yeast pheromone signaling pathway. Mol Cell Proteomics 4:310
4. Wolf-Yadlin A, Hautaniemi S, Lauffenburger A, White M (2007) Multiple reaction monitoring for robust quantitative proteomic analysis of cellular signaling networks. Proc Natl Acad Sci USA 104:5860
5. Cantin T, Venable D, Cociorva D, Yates R (2006) Iii quantitative phosphoproteomic analysis of the tumor necrosis factor pathway. J Proteome Res 5:127
6. Beausoleil A, Jedrychowski M, Schwartz D, Elias E, Villen J, Li J, Cohn A, Cantley C, Gygi P (2004) Large-scale characterization of HeLa cell nuclear phosphoproteins. Proc Natl Acad Sci USA 101:12130
7. Olsen V, Blagoev B, Gnad F, Macek B, Kumar C, Mortensen P, Mann M (2006) Global, in vivo, and site-specific phosphorylation dynamics in signaling networks. Cell 127:635
8. Hoffert J, Pisitkun T, Wang G, Shen R, Knepper M (2006) Quantitative phosphoproteomics of vasopressin-sensitive renal cells: regulation of aquaporin-2 phosphorylation at two sites. Proc Natl Acad Sci USA
9. Saeed F, Pisitkun T, Hoffert JD, Wang G, Gucek M, Knepper MA (2012) An efficient dynamic programming algorithm for phosphorylation site assignment of large-scale mass spectrometry data. In: 2012 IEEE international conference on bioinformatics and biomedicine workshops (BIBMW), vol 11. IEEE, BioMed Central Ltd, pp 618–625
10. Pisitkun T, Shen R-F, Knepper MA (2004) Identification and proteomic profiling of exosomes in human urine. Proc Natl Acad Sci USA 101(36):13368–13373
11. Zhao B, Pisitkun T, Hoffert JD, Knepper MA, Saeed F (2012) CP hos: a program to calculate and visualize evolutionarily conserved functional phosphorylation sites. Proteomics 12(22):3299–3303
12. Linnet K (2013) Toxicological screening and quantitation using liquid chromatography/time-of-flight mass spectrometry. J Forensic Sci Criminol 1(1):1
13. Gika HG, Theodoridis GA, Plumb RS, Wilson ID (2014) Current practice of liquid chromatography-mass spectrometry in metabolomics and metabonomics. J Pharm Biomed Anal 87:12–25
14. Hoffert J, Pisitkun T, Wang G, Shen F, Knepper M (2006) Quantitative phosphoproteomics of vasopressin-sensitive renal cells: regulation of aquaporin-2 phosphorylation at two sites. Proc Natl Acad Sci USA 103(18):7159–7164
15. Li X, Gerber SA, Rudner AD, Beausoleil SA, Haas W, Elias JE, Gygi SP (2007) Large-scale phosphorylation analysis of alpha-factor-arrested saccharomyces cerevisiae. J Proteome Res 6(3):1190–1197. http://www.biomedsearch.com/nih/Large-scale-phosphorylation-analysis-alpha/17330950.html

16. Gruhler A, Olsen JV, Mohammed S, Mortensen P, Faergeman NJ, Mann M, Jensen ON (2005) Quantitative phosphoproteomics applied to the yeast pheromone signaling pathway. Mol Cell Proteomics 4(3):310–327. https://doi.org/10.1074/mcp.M400219-MCP200
17. Doerr A (2014) Dia mass spectrometry. Nat Methods 12(1):35
18. Gillet LC, Navarro P, Tate S, Röst H, Selevsek N, Reiter L, Bonner R, Aebersold R Targeted data extraction of the MS/MS spectra generated by data-independent acquisition: a new concept for consistent and accurate proteome analysis. Mol Cell Proteomics 11(6). https://www.mcponline.org/content/11/6/O111.016717.full.pdf, https://doi.org/10.1074/mcp.O111.016717
19. Egertson JD, Kuehn A, Merrihew GE, Bateman NW, MacLean BX, Ting YS, Canterbury JD, Marsh DM, Kellmann M, Zabrouskov V, et al (2013) Multiplexed MS/MS for improved data-independent acquisition. Nat Methods 10(8):744
20. Bern M, Finney G, Hoopmann MR, Merrihew G, Toth MJ, MacCoss MJ (2009) Deconvolution of mixture spectra from ion-trap data-independent-acquisition tandem mass spectrometry. Anal Chem 82(3):833–841
21. Tsou C-C, Avtonomov D, Larsen B, Tucholska M, Choi H, Gingras A-C, Nesvizhskii AI (2015) DIA-Umpire: comprehensive computational framework for data-independent acquisition proteomics. Nat Methods 12(3):258
22. Li Y, Zhong C-Q, Xu X, Cai S, Wu X, Zhang Y, Chen J, Shi J, Lin S, Han J (2015) Group-DIA: analyzing multiple data-independent acquisition mass spectrometry data files. Nat Methods 12(12):1105
23. Wang J, Tucholska M, Knight JD, Lambert J-P, Tate S, Larsen B, Gingras A-C, Bandeira N (2015) MSPLIT-DIA: sensitive peptide identification for data-independent acquisition. Nat Methods 12(12):1106
24. Meyer JG, Mukkamalla S, Steen H, Nesvizhskii AI, Gibson BW, Schilling B (2017) PIQED: automated identification and quantification of protein modifications from DIA-MS data. Nat Methods 14(7):646
25. Tran NH, Qiao R, Xin L, Chen X, Liu C, Zhang X, Shan B, Ghodsi A, Li M (2019) Deep learning enables de novo peptide sequencing from data-independent-acquisition mass spectrometry. Nat Methods 16(1):63–66
26. Röst HL, Rosenberger G, Navarro P, Gillet L, Miladinović SM, Schubert OT, Wolski W, Collins BC, Malmström J, Malmström L, et al (2014) Openswath enables automated, targeted analysis of data-independent acquisition MS data. Nat Biotechnol 32(3):219
27. Bilbao A, Varesio E, Luban J, Strambio-De-Castillia C, Hopfgartner G, Müller M, Lisacek F (2015) Processing strategies and software solutions for data-independent acquisition in mass spectrometry. Proteomics 15(5–6):964–980
28. Castillo S, Gopalacharyulu P, Yetukuri L, Orešič M (2011) Algorithms and tools for the preprocessing of LC-MS metabolomics data. Chemom Intell Lab Syst 108(1):23–32
29. Awan MG, Saeed F (2016) MS-Reduce: an ultrafast technique for reduction of big mass spectrometry data for high-throughput processing. Bioinformatics 32(10):1518–1526
30. Mujezinovic N, Raidl G, Hutchins JR, Peters J-M, Mechtler K, Eisenhaber F (2006) Cleaning of raw peptide MS/MS spectra: improved protein identification following deconvolution of multiply charged peaks, isotope clusters, and removal of background noise. Proteomics 6(19):5117–5131
31. Ding J, Shi J, Poirier GG, Wu F-X (2009) A novel approach to denoising ion trap tandem mass spectra. Proteome Sci 7(1):9
32. Pluskal T, Castillo S, Villar-Briones A, Orešič M (2010) Mzmine 2: modular framework for processing, visualizing, and analyzing mass spectrometry-based molecular profile data. BMC Bioinform 11(1):395
33. Xia J, Psychogios N, Young N, Wishart DS (2009) Metaboanalyst: a web server for metabolomic data analysis and interpretation. Nucl Acids Res 37(suppl_2):W652–W660
34. Frank AM, Bandeira N, Shen Z, Tanner S, Briggs SP, Smith RD, Pevzner PA (2007) Clustering millions of tandem mass spectra. J Proteome Res 7(01):113–122

35. Yuan ZF, Liu C, Wang HP, Sun RX, Fu Y, Zhang JF, Wang LH, Chi H, Li Y, Xiu LY, et al pParse: a method for accurate determination of monoisotopic peaks in high-resolution mass spectra. Proteomics 12(2):226–235
36. Beer I, Barnea E, Ziv T, Admon A (2004) Improving large-scale proteomics by clustering of mass spectrometry data. Proteomics 4(4):950–960
37. Saeed F, Hoffert JD, Knepper MA (2014) Cams-rs: clustering algorithm for large-scale mass spectrometry data using restricted search space and intelligent random sampling. IEEE/ACM Trans Comput Biol Bioinform (TCBB) 11(1):128–141
38. The M, Lukas K (2016) Maracluster: a fragment rarity metric for clustering fragment spectra in shotgun proteomics. J Proteome Res 15(3):713–720
39. Nesvizhskii AI (2010) A survey of computational methods and error rate estimation procedures for peptide and protein identification in shotgun proteomics. J Proteomics 73(11):2092–2123
40. Frank A, Pevzner P (2005) PepNovo: de novo peptide sequencing via probabilistic network modeling. Anal Chem 77(4):964–973
41. Chi H, Sun R-X, Yang B, Song C-Q, Wang L-H, Liu C, Fu Y, Yuan Z-F, Wang H-P, He S-M et al (2010) pNovo: de novo peptide sequencing and identification using HCD spectra. J Proteome Res 9(5):2713–2724
42. Yang H, Chi H, Zhou W-J, Zeng W-F, He K, Liu C, Sun R-X, He S-M (2017) Open-pNovo: de novo peptide sequencing with thousands of protein modifications. J Proteome Res 16(2):645–654
43. Tran NH, Zhang X, Xin L, Shan B, Li M (2017) De novo peptide sequencing by deep learning. Proc Natl Acad Sci 114(31):8247–8252
44. Ma B, Zhang K, Hendrie C, Liang C, Li M, Doherty-Kirby A, Lajoie G (2003) Peaks: powerful software for peptide de novo sequencing by tandem mass spectrometry. Rapid Commun Mass Spectrom 17(20):2337–2342
45. Taylor JA, Johnson RS (2001) Implementation and uses of automated de novo peptide sequencing by tandem mass spectrometry. Anal Chem 73(11):2594–2604
46. Bandeira N (2007) Spectral networks: a new approach to de novo discovery of protein sequences and posttranslational modifications. Biotechniques 42(6):687–695
47. Grossmann J, Roos FF, Cieliebak M, Lipták Z, Mathis LK, Müller M, Gruissem W, Baginsky S (2005) Audens: a tool for automated peptide de novo sequencing. J Proteome Res 4(5):1768–1774
48. Mo L, Dutta D, Wan Y, Chen T (2007) Msnovo: a dynamic programming algorithm for de novo peptide sequencing via tandem mass spectrometry. Anal Chem 79(13):4870–4878
49. Fernandez-de-Cossio J, Gonzalez J, Satomi Y, Shima T, Okumura N, Besada V, Betancourt L, Padron G, Shimonishi Y, Takao T (2000) Automated interpretation of low-energy collision-induced dissociation spectra by SeqMS, a software aid for de novo sequencing by tandem mass spectrometry. ELECTROPHORESIS: An Int J 21(9):1694–1699
50. Jagannath S, Sabareesh V (2007) Peptide fragment ion analyser (PFIA): a simple and versatile tool for the interpretation of tandem mass spectrometric data and de novo sequencing of peptides. Rapid Commun Mass Spectrom: Int J Devoted Rapid Dissem Up-To–Minute Res Mass SpectrometryRapid Commun Mass Spectrom: Int J Devoted Rapid Dissem Up-to-the-Minute Res Mass Spectrom 21(18):3033–3038
51. Fischer B, Roth V, Roos F, Grossmann J, Baginsky S, Widmayer P, Gruissem W, Buhmann JM (2005) NovoHMM: a hidden Markov model for de novo peptide sequencing. Anal Chem 77(22):7265–7273
52. Lam H, Deutsch E, Eddes J, Eng J, King N, Yang S, Roth J, Kilpatrick L, Neta P, Stein S, et al (2006) Spectrast: an open-source MS/MS spectramatching library search tool for targeted proteomics. In: Poster at 54th ASMS conference on mass spectrometry, pp 1–10
53. Bittremieux W, Meysman P, Noble WS, Laukens K (2018) Fast open modification spectral library searching through approximate nearest neighbor indexing. J Proteome Res 17(10):3463–3474

54. Ye D, Fu Y, Sun R-X, Wang H-P, Yuan Z-F, Chi H, He S-M (2010) Open MS/MS spectral library search to identify unanticipated post-translational modifications and increase spectral identification rate. Bioinformatics 26(12):i399–i406
55. Dasari S, Chambers MC, Martinez MA, Carpenter KL, Ham A-JL, Vega-Montoto LJ, Tabb DL (2012) Pepitome: evaluating improved spectral library search for identification complementarity and quality assessment. J Proteome Res 11(3):1686–1695
56. Griss J (2016) Spectral library searching in proteomics. Proteomics 16(5):729–740
57. Lam H, Aebersold R (2010) Spectral library searching for peptide identification via tandem MS. In: Proteome bioinformatics. Springer, pp 95–103
58. Hummel J, Niemann M, Wienkoop S, Schulze W, Steinhauser D, Selbig J, Walther D, Weckwerth W (2007) ProMEX: a mass spectral reference database for proteins and protein phosphorylation sites. BMC Bioinform 8(1):216
59. Wu X, Tseng C-W, Edwards N (2007) HMMatch: peptide identification by spectral matching of tandem mass spectra using hidden Markov models. J Comput Biol 14(8):1025–1043
60. Wu Z, Lajoie G, Ma B (2008) MSDash: mass spectrometry database and search. In: Computational systems bioinformatics: vol 7. World Scientific, pp 63–71
61. Ahrne E, Nikitin F, Lisacek F, Muller M (2011) QuickMod: a tool for open modification spectrum library searches. J Proteome Res 10(7):2913–2921
62. Geer LY, Markey SP, Kowalak JA, Wagner L, Xu M, Maynard DM, Yang X, Shi W, Bryant SH (2004) Open mass spectrometry search algorithm. J Proteome Res 3(5):958–964
63. Horlacher O, Lisacek F, Markus M (2015) Mining large scale tandem mass spectrometry data for protein modifications using spectral libraries. J Proteome Res 15(3):721–731
64. Kong AT, Leprevost FV, Avtonomov DM, Mellacheruvu D, Nesvizhskii AI (2017) MSFragger: ultrafast and comprehensive peptide identification in mass spectrometry-based proteomics. Nat Methods 14(5):513
65. Tiwary S, Levy R, Gutenbrunner P, Soto FS, Palaniappan KK, Deming L, Berndl M, Brant A, Cimermancic P, Cox J (2019) High-quality MS/MS spectrum prediction for data-dependent and data-independent acquisition data analysis. Nat Methods 16(6):519
66. Gessulat S, Schmidt T, Zolg DP, Samaras P, Schnatbaum K, Zerweck J, Knaute T, Rechenberger J, Delanghe B, Huhmer A et al (2019) Prosit: proteome-wide prediction of peptide tandem mass spectra by deep learning. Nat Methods 16(6):509
67. Zhou X-X, Zeng W-F, Chi H, Luo C, Liu C, Zhan J, He S-M, Zhang Z (2017) pDeep: predicting MS/MS spectra of peptides with deep learning. Anal Chem 89(23):12690–12697
68. Gabriels R, Martens L, Degroeve S (2019) Updated MS^{2}pip web server delivers fast and accurate MS^2 peak intensity prediction for multiple fragmentation methods, instruments and labeling techniques. Nucleic Acids Res 47(W1):W295–W299
69. Eng JK, McCormack AL, Yates JR (1994) An approach to correlate tandem mass spectral data of peptides with amino acid sequences in a protein database. J Am Soc Mass Spectrom 5(11):976–989
70. David M, Fertin G, Rogniaux H, Tessier D (2017) SpecOMS: a full open modification search method performing all-to-all spectra comparisons within minutes. J Proteome Res 16(8):3030–3038
71. Chi H, Liu C, Yang H, Zeng W-F, Wu L, Zhou W-J, Wang R-M, Niu X-N, Ding Y-H, Zhang Y, et al (2018) Comprehensive identification of peptides in tandem mass spectra using an efficient open search engine. Nat Biotechnol 36(11):1059
72. Cox J, Neuhauser N, Michalski A, Scheltema RA, Olsen JV, Mann M (2011) Andromeda: a peptide search engine integrated into the Maxquant environment. J Proteome Res 10(4):1794–1805
73. Craig R, Beavis RC (2004) Tandem: matching proteins with tandem mass spectra. Bioinformatics 20(9):1466–1467
74. Park CY, Klammer AA, Kall L, MacCoss MJ, Noble WS (2008) Rapid and accurate peptide identification from tandem mass spectra. J Proteome Res 7(7):3022–3027
75. Diament BJ, Noble WS (2011) Faster SEQUEST searching for peptide identification from tandem mass spectra. J Proteome Res 10(9):3871–3879

76. Eng JK, Fischer B, Grossmann J, MacCoss MJ (2008) A fast SEQUEST cross correlation algorithm. J Proteome Res 7(10):4598–4602
77. Devabhaktuni A, Lin S, Zhang L, Swaminathan K, Gonzalez CG, Olsson N, Pearlman SM, Rawson K, Elias JE (2019) Taggraph reveals vast protein modification landscapes from large tandem mass spectrometry datasets. Nat Biotechnol 37(4):1
78. Zhang J, Xin L, Shan B, Chen W, Xie M, Yuen D, Zhang W, Zhang Z, Lajoie GA, Ma B (2012) Peaks DB: de novo sequencing assisted database search for sensitive and accurate peptide identification. Mol Cell Proteomics 11(4):M111-010587
79. Wang X, Li Y, Wu Z, Wang H, Tan H, Peng J (2014) Jump: a tag-based database search tool for peptide identification with high sensitivity and accuracy. Mol Cell Proteomics 13(12):3663–3673
80. Bern M, Cai Y, Goldberg D (2007) Lookup peaks: a hybrid of de novo sequencing and database search for protein identification by tandem mass spectrometry. Anal Chem 79(4):1393–1400
81. Chi H, He K, Yang B, Chen Z, Sun R-X, Fan S-B, Zhang K, Liu C, Yuan Z-F, Wang Q-H et al (2015) pFind-Alioth: a novel unrestricted database search algorithm to improve the interpretation of high-resolution MS/MS data. J Proteomics 125:89–97
82. Li Y, Chi H, Wang L-H, Wang H-P, Fu Y, Yuan Z-F, Li S-J, Liu Y-S, Sun R-X, Zeng R et al (2010) Speeding up tandem mass spectrometry based database searching by peptide and spectrum indexing. Rapid Commun Mass Spectrom 24(6):807–814
83. Mann M, Wilm M (1994) Error-tolerant identification of peptides in sequence databases by peptide sequence tags. Anal Chem 66(24):4390–4399
84. Tabb DL, Saraf A, Yates JR (2003) GutenTag: high-throughput sequence tagging via an empirically derived fragmentation model. Anal Chem 75(23):6415–6421
85. Dasari S, Chambers MC, Codreanu SG, Liebler DC, Collins BC, Pennington SR, Gallagher WM, Tabb DL (2011) Sequence tagging reveals unexpected modifications in toxicoproteomics. Chem Res Toxicol 24(2):204–216
86. Dasari S, Chambers MC, Slebos RJ, Zimmerman LJ, Ham A-JL, Tabb DL (2010) Tagrecon: high-throughput mutation identification through sequence tagging. J Proteome Res 9(4):1716–1726
87. Searle BC, Dasari S, Wilmarth PA, Turner M, Reddy AP, David LL, Nagalla SR (2005) Identification of protein modifications using MS/MS de novo sequencing and the opensea alignment algorithm. J Proteome Res 4(2):546–554
88. Tanner S, Shu H, Frank A, Wang L-C, Zandi E, Mumby M, Pevzner PA, Bafna V (2005) Inspect: identification of posttranslationally modified peptides from tandem mass spectra. Anal Chem 77(14):4626–4639
89. Tanner S, Pevzner PA, Bafna V (2006) Unrestrictive identification of post-translational modifications through peptide mass spectrometry. Nat Protoc 1(1):67
90. Chick JM, Kolippakkam D, Nusinow DP, Zhai B, Rad R, Huttlin EL, Gygi SP (2015) A mass-tolerant database search identifies a large proportion of unassigned spectra in shotgun proteomics as modified peptides. Nat Biotechnol 33(7):743
91. Griss J, Perez-Riverol Y, Lewis S, Tabb DL, Dianes JA, del Toro N, Rurik M, Walzer M, Kohlbacher O, Hermjakob H, et al (2016) Recognizing millions of consistently unidentified spectra across hundreds of shotgun proteomics datasets. Nat Methods 13(8):651
92. Lundgren DH, Han DK, Eng JK (2005) Protein identification using turbosequest. Curr Protoc Bioinform 10(1):13–3
93. Kim S, Pevzner PA (2014) Ms-gf+ makes progress towards a universal database search tool for proteomics. Nat Commun 5(1):5277
94. Clauser KR, Baker P, Burlingame AL (1999) Role of accurate mass measurement (±10 ppm) in protein identification strategies employing MS or MS/MS and database searching. Anal Chem 71(14):2871–2882
95. Perkins DN, Pappin DJ, Creasy DM, Cottrell JS (1999) Probability-based protein identification by searching sequence databases using mass spectrometry data. ELECTROPHORESIS: An Int J 20(18):3551–3567

96. Käll L, Canterbury JD, Weston J, Noble WS, MacCoss MJ (2007) Semi-supervised learning for peptide identification from shotgun proteomics datasets. Nat Methods 4(11):923–925
97. Brosch M, Yu L, Hubbard T, Choudhary J (2009) Accurate and sensitive peptide identification with mascot percolator. J Proteome Res 8(6):3176–3181
98. Spivak M, Weston J, Bottou L, Käll L, Noble WS (2009) Improvements to the percolator algorithm for peptide identification from shotgun proteomics data sets. J Proteome Res 8(7):3737–3745
99. Keller A, Nesvizhskii AI, Kolker E, Aebersold R (2002) Empirical statistical model to estimate the accuracy of peptide identifications made by MS/MS and database search. Anal Chem 74(20):5383–5392
100. Shteynberg D, Deutsch EW, Lam H, Eng JK, Sun Z, Tasman N, Mendoza L, Moritz RL, Aebersold R, Nesvizhskii AI (2011) iProphet: multi-level integrative analysis of shotgun proteomic data improves peptide and protein identification rates and error estimates. Mol Cell Proteomics 10(12):M111-007690
101. Nesvizhskii AI (2014) Proteogenomics: concepts, applications and computational strategies. Nat Methods 11(11):1114–1125
102. Zhu Y, Orre LM, Johansson HJ, Huss M, Boekel J, Vesterlund M, Fernandez-Woodbridge A, Branca RM, Lehtiö J (2018) Discovery of coding regions in the human genome by integrated proteogenomics analysis workflow. Nat Commun 9(1):903
103. Risk BA, Spitzer WJ, Giddings MC (2013) Peppy: proteogenomic search software. J Proteome Res 12(6):3019–3025
104. Jagtap PD, Johnson JE, Onsongo G, Sadler FW, Murray K, Wang Y, Shenykman GM, Bandhakavi S, Smith LM, Griffin TJ (2014) Flexible and accessible workflows for improved proteogenomic analysis using the galaxy framework. J Proteome Res 13(12):5898–5908
105. Nagaraj SH, Waddell N, Madugundu AK, Wood S, Jones A, Mandyam RA, Nones K, Pearson JV, Grimmond SM (2015) PGTools: a software suite for proteogenomic data analysis and visualization. J Proteome Res 14(5):2255–2266
106. Castellana NE, Shen Z, He Y, Walley JW, Briggs SP, Bafna V (2014) An automated proteogenomic method uses mass spectrometry to reveal novel genes in Zea mays. Mol Cell Proteomics 13(1):157–167

Chapter 3
Existing HPC Methods and the Communication Lower Bounds for Distributed-Memory Computations for Mass Spectrometry-Based Omics Data

Fahad Saeed and Muhammad Haseeb

Mass spectrometry (MS)-based omics data analysis requires substantial time and resources which has necessitated the need for high-performance computing (HPC) methods. Few parallel algorithms have been proposed, designed, and developed when the amount of data that needed to be processed was smaller in scale, i.e. only a few PTM were of interest and would satisfy when only a shorter theoretical database was needed for computations. However, with the increase in the amount of data that needs processing due to the expansion of theoretical databases due to a large number of PTMs that are needed for various system biology problems; the current HPC methods are not sufficiently scalable with the increasing size of the database or the processors. However, all of this is anecdotal and more evidence is needed to quantify how good or bad the current HPC methods were. We also wanted to know if there was a more scalable algorithm that could be possible and this is the objective of this chapter.

3.1 Introduction

Numerical algorithms have tried to improved the efficiency of *arithmetic operations* for several decades now [1]. Moore's law increased the size and the efficiency of the computer chips which has resulted in increasing the arithmetic operations faster. Therefore, many of the current numerical algorithms are bottlenecked by the *communication* costs of the data movement in contrast to the speed of the arithmetic operations. This communication cost can be due to data movement between different levels of the hierarchy of memory or it could be due to communication of the data between different processing units and different distributed-memory processors

Some parts of this chapter may have appeared in [25].

F. Saeed and M. Haseeb, *High-Performance Algorithms for Mass Spectrometry-Based Omics*, Computational Biology 34,
https://doi.org/10.1007/978-3-031-01960-9_3

connected via a network. Overwhelming studies have shown that the cost of moving the data exceeds the cost of arithmetic operations in most of the numerical algorithms [1–3]. New generation of architectures that include multicore, manycore, and CPU-GPU architectures will increase the gap between the communication costs and the computation costs, i.e. communication costs will become *more* significant over time [4, 5]. This trend is observed both for serial as well as parallel algorithms [1].

To increase the efficiency of data analysis workflow, significant investment of time and resources has poured in and resulted in efficient *numerical methods* [6–10]. All of these methods have considered arithmetic operations as the sole metric for efficiency. MS-based experiments can now generate larger and more complex (multiple species, non-model species, microbiome) data sets that need more complex search engine mechanisms including but not limited to searching the data sets using a large number of PTMs. This is essential for large-scale systems biology studies that can lead to high-impact proteomics, meta-proteomics, and microbiome studies directly related to human disease and health.

Although the development of computational methods for MS data analysis is an active area of research, the overall objective of these research studies and workflows is to increase the computational efficiency of the methods. Database search mechanism is the most widely used data processing pipeline which works by matching MS spectra with a database of theoretical spectra generated in silico. These theoretical databases can grow several times larger than the original proteome and is highly dependent on the search parameters used and can be larger for meta-proteomics or non-model studies [11, 12]. This data can easily reach terabyte level leading to intractable runtimes for the current genre of serial numerical algorithms. Since serial algorithms can easily be overwhelmed when the theoretical data is significantly larger than the available memory on the system, researchers have retort to developing HPC algorithms [13, 14, 14–18] for these MS-based omics workflows. Similar to serial algorithms, the HPC methods have also been designed and implemented as a way to speed up the arithmetic scoring part with no effort towards minimizing the communication costs. These excessive communications costs lead to the performance of serial [6–10], and HPC methods that are less than stellar which may have dampened the enthusiasm of proteomics practitioners due to excessive processing times for these workflows.

There is sufficient anecdotal evidence for systems biology practitioners and proteomics domain experts. However, how do we quantify if the reported methods are adequately scalable and if there are algorithms that *could* be proposed for a better performance. Therefore, we set out to ask the following two questions that could help us answer these questions: (1) What are the lower bounds of the currently reported state-of-the-art HPC methods; (2) If these lower bounds can be improved with the introduction of new parallel algorithms and if that is possible. The bounds will be similar for serial algorithms subject to architecture-specific communication costs. Our anecdotal evidence also suggested that not a lot of performance was gained when the number of processing units is increased for the existing parallel algorithms. We needed to investigate why this is the case, and if in fact the communication costs are the bottleneck for these parallel algorithms and that this performance is not due to a specific library or an artifact of the architecture.

In this chapter, we prove that the efficiency of these parallel algorithms is bottlenecked by communication costs. We also show that the theoretical lower bounds that are possible are not achieved by any of the existing state-of-the-art HPC methods. At the end of the chapter, we summarize all of the scalability data that we had collected from published reports to observe and demonstrate that theoretical lower bounds are in agreement with the empirical observations.

Attaining the lower bounds that we prove in this chapter would require a significant redesign of parallel algorithm which can process MS-based omics data sets. In contrast to existing practices of including few thread, or OPENMP parameters in parallel code, these redesigns would have to exploit numerical properties, MS data compressive analytics , decomposition of data on distributed architectures, and focus on developing metrics that minimize the communication costs associated with these parallel algorithms.

The rest of the chapter is organized as follows. In Sect. 3.2, we formulate the communication models to analyze the current parallel algorithms. This is followed by Sect. 3.3, we briefly introduce the reader to proteomics workflows and summarize (and reduce) the parallelization strategy that is used by current HPC methods. Thereafter, in Sect. 3.4, we provide theoretical communications bounds, the computation bounds, and the overall runtime bounds of the existing, and (possible but not yet discovered) communication-optimal parallel algorithms. The next Sect. 3.5 provides the meta-data analysis of published results till date and a short description of how this meta-data analysis is in agreement with our theoretical bounds that we just proved. The chapter concludes with a useful discussion about the current and future research efforts that are needed in this domain.

3.2 Communication Model

Design of parallel algorithms for big data problem requires that the load is balanced on all processing units, and the communication costs are minimal as compared to the computation costs [19]. The parallel algorithms that are proposed for MS-based omics can be modeled in the following way: The two costs associated with communication are when the parallel algorithm sends n words from one processor to another via the network. In this case, the words are first packed in contiguous blocks or memory and are known as *message* in parallel computing or MPI jargon. Once the message is packed, it is then sent to the other processor as the parallel algorithmic might dictate. The times it takes to assemble, pack, and transmit the data is an overhead that is fixed depending on the computer infrastructure, and this *latency cost* is denoted by α in our model. *Bandwidth cost* is the time it needs to transmit the n words and is denoted by βn. Therefore, the cost to send one message of n words is denoted by $\alpha + \beta n$ and the time to send S messages containing a total of W words can be denoted by $\alpha S + \beta W$. Also assume that γ denotes the time to complete one arithmetic computation, and the total number of computations is denoted by F. Summation of all of these terms is equal to $\alpha S + \beta W + \gamma F$ and it is easier to see that $\alpha >> \beta >> \gamma$. Therefore, parallel algorithms that can minimize both the bandwidth and latency will lead to scalable algorithms. The communication model that we have just presented is used for developing communication-minimizing parallel algorithms for a plethora of linear-algebraic computations.

3.2.1 Sequential Computer

For two levels of memory hierarchy, the model $\alpha S + \beta W + \gamma F$, would suffice. The model can be extended if there are more levels of hierarchy available in a specific system where the data moved to/from different levels have to incorporate into the model.

3.2.2 Parallel Computer

Our model that uses $\alpha S + \beta W + \gamma F$ would be sufficient to provide the communication costs that may be associated with one node of a parallel computing architecture. For our purposes, we will assume a single homogeneous supercomputer which is a distributed-memory architecture. The advantage of this assumption is that the lower bounds calculated on one processor is enough to get lower bounds on the whole algorithm for completing the same tasks. The upper bounds (time for the whole algorithm) will be the summation of all the terms ordered with dependencies considering the critical paths which maximize those costs, i.e. If the parallel algorithm can allow an overlap between communication and computation, then the expression can be replaced with $max(\alpha S + \beta W, \gamma F)$ or $max(\alpha S, \beta W, \gamma F)$ but does not affect the asymptotic relations. Lastly, we will define *Communication-avoiding*, or communication-optimal algorithm for methods that can attain asymptotically lower bounds of communication for a given architecture.

3.3 MS Database Proteomics, Proteogenomics, and Meta-Proteomics Search

To prove our lower bounds in the most generalized form applicable to a wide variety of MS-based omics workflows, we will first define *database-search* strategy used Mass spectrometry-based proteomics in its simplest and bare bone form without considering the systems biology objectives or method-specific optimizations. Peptide identification is commonly accomplished by performing a database search where the experimentally obtained MS data spectra are compared (and scored) with a theo-

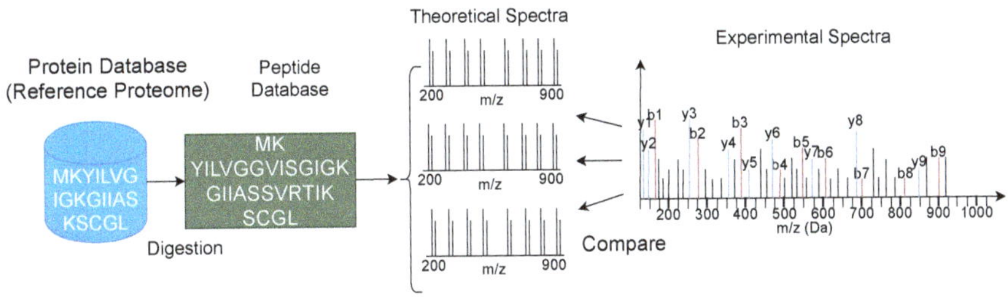

Fig. 3.1 A high-level overview of the MS-based proteomics data analysis that leads to spectra-to-peptide deductions

retically predicted spectra-database [10]. This theoretical spectra-database (or theoretical database) is obtained by first performing a *in silico* digestion of the proteome sequence database into peptides. These peptides are then used to predict the MS/MS spectra, and its possible (modified) variants,[1] including post-translation modifications (PTMs), fragmentation types, and the associated mass windows. The matching and scoring of the spectra[2] with the theoretical database is known as peptide-to-spectrum match (PSM) computations, and a high-level overview is shown in Fig. 3.1.

3.3.1 Generalized Parallel Computing Strategy

All parallel algorithms (except for HiCOPS [20]) have been proposed as numerical algorithms that have been designed to improve the computational efficiency of the parallel methods. All of the existing HPC methods work in the following way: Assume that there are N spectra that need to be processed using p processors. All of these methods divide N spectra among p processors such that N/p spectra are processed on each processing unit. The underlying assumption is that either database is divided (and replicated equally among processing units) or that (smaller) FASTA database is communicated which is then expanded on each machine. In either case, database D is either communicated or expanded. After this initial communication, these HPC algorithms operate by executing a serial method (such as XTandem) on each node in parallel. Completion of the computations triggers each processor to transmit the result back to the master node. A sketch of this is illustrated in Algorithm 1. Few assumptions that are made when developing these HPC methods include equal load for each spectra computation with minimal communication costs and assumption that load-balanced system can be achieved by dividing the spectra into equal numbers. We now show in this chapter that the assumptions are a source of major bottleneck for these parallel algorithms.

[1] Theoretical database generally will refer to the in-silico predicted MS/MS spectra from these peptides.

[2] Spectra will generally refer to spectra obtained experimentally unless otherwise stated.

Algorithm 1: General HPC strategy that is used by Parallel Methods for MS based Proteomics data

Result: Each Spectra is assigned to a peptide

1 **while** *Spectra need peptide deduction* **do**

1. Take a species-specific protein database; and expand it to a theoretical database D using search parameters;
2. Database D is copied whole on each of the P processors;
3. The spectra set S that needs to be processed are divided in S/P parts;
4. S/P spectra are processed on each of the processor in parallel;
5. The results are accumulated using MPI-gather or similar operation;

2 **end**

3.4 Communication Lower Bounds

We formulate the problem in terms of matrix calculations and prove both the communication and computation bounds that are exhibited by these HPC methods.

Definition 3.1 Theoretical database is produced in-silico using peptides that are obtained from the proteome database. This theoretical database is expanded using the search parameters specific to the PTMs. Assume that this database is $m \times n$ matrix D where m presented the number of theoretical spectra entries, and n presents the average length of the entries. The entries of matrix D can be access using i and j indexes where $(0 \leq i < n)$ and $(0 \leq j < m)$. Then rows of D can be accessed using $D(0, j)$, $D(1, j)$, and so on.

Definition 3.2 Let the *set* of spectra $s_0, s_1, \ldots, s_{(q-1)}$ that needs to be processed can be presented by a matrix S $q \times r$, where q represents the number of spectra and r represents the average length of the spectra. The entries of matrix S can be access using i and j indexes, where $(0 \leq i < r)$ and $(0 \leq j < q)$. Then rows of S can be access using $S(0, j)$, $S(1, j)$, and so on.

Matrix D, matrix S, and the peptides that are deduced are shown in Fig. 3.2.

Definition 3.3 The parallel architecture is a distributed-memory processor with M fast memory associated with a single-core processor. We will assume that all processors are connected to each other via a network.

Lemma 3.1 *There are three communication rounds that will take place to complete the computations as shown in Algorithm 1 for the existing parallel algorithms.*

Proof One communication round is the distribution of the database on each of the processors. The second communication round is the distribution of q/p spectra on each processor. The third communication round is completed when processing on each compute node is completed and q/p messages are accumulated on a single master machine.

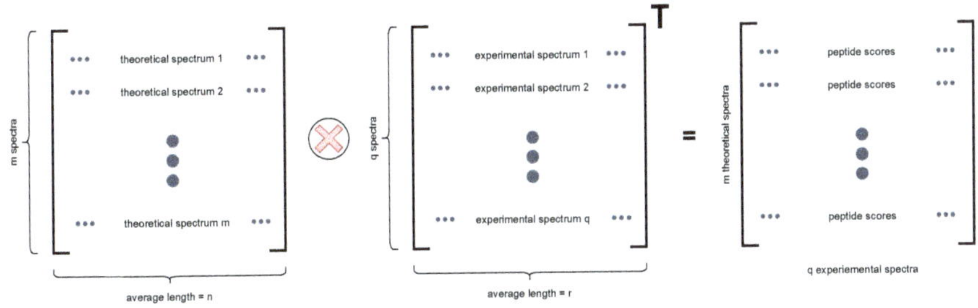

Fig. 3.2 A schematic of the matrix D (that represents the theoretical spectra), matrix S (that represents the experimental spectra), and matrix that holds the peptide that is deduced

Theorem 3.1 *The total words that are communicated using three rounds listed above are equal to $\Omega(mn)$ for existing HPC strategies.*

Proof The total words communicated on each processor is equal to $|D| + \frac{|S|}{p} + \frac{|S|}{p}$. Here, it is easy to see that $|D| = (m \times n)$. Further, $\frac{|S|}{p}$ is going to be equal to the words that are communicated from the spectra set i.e. $\frac{q \times r}{p}$. The final communication round is when the peptides are deduced for each spectrum and accumulated on a processor. The words that will be communicated are equal to $\frac{q \times r}{p}$. r is assumed to be the case where the spectra peaks are equal to the peptide length. Then the total number of words that are communicated is equal to $(m \times n) + \frac{q \times r}{p} + \frac{q \times r}{p} = (m \times n) + 2\frac{q \times r}{p}$. Therefore, the words communicated is $\Omega(mn + 2r\frac{q}{p})$.

Theorem 3.2 *The computational costs of dot product like scoring that is performed for peptide-to-spectrum match for each processor is equal to $F = \frac{qm(2n-1)}{p}$.*

Proof Each scalar dot product (called score) will work on one array from the database D and one array from the spectra S. On processor P0 which contains the whole matrix D and subset of matrix from S, a score is calculated for $D(0, i)$ $0 \leq i \leq n$ and $S(0, j)$ $0 \leq j \leq r$. This will require n multiplications and $(n - 1)$ additions. Since this has to be done for all entries of database D, it will require $m \times (2n - 1)$ computation for a single spectrum. It is obvious that the number of spectra on each processor is q/p. This implies that the words that need to be processed on each processor are $\frac{qm(2n-1)}{p}$.

Theorem 3.3 *The lower bound of bandwidth communication for database spectra to peptide match is $W = \Omega(\frac{m}{p})$ for any configuration of database or spectra in which dot-product scores are performed for matching.*

Proof The lower bound of communication possible is equal to $\Omega(\#of\,Flops/\sqrt{M})$. The computations required for dot product like routines is $O(\frac{qm(2n)}{p})$ as proved in our earlier theorem. The size of the fast memory is assumed to contain both the database, the spectra that needs to be searched and the result of the scoring. Therefore, $\sqrt{M} \geq mn + \frac{2qr}{p}$. Then the equation $\Omega(\frac{2qm}{p \times (mn + \frac{2qr}{p})})$. Our earlier assumption that $n \approx r$ and q can be approaching m is applicable here without losing generality.

This gives us $\frac{m^2 n}{pmn+2qn}$ which is equivalent to $\frac{m^2}{pm+2q}$. For M which can contain the database, the spectra, and the results; As before for $q \approx m$ proves that lower bound of communication which can be reached is equal to $\Omega(\frac{m}{p})$.

Theorem 3.4 *We prove that the lower bound on the Latency cost* $L = \Omega(\frac{2}{mpn^2})$

Proof A lower bandwidth bound on the bandwidth cost W gives us a lower bound on the latency cost L. Assume that the largest message by a given architecture is m_{max}, then it is clear that $L \geq W/m_{max}$ since no message can be larger than the memory. Therefore, we get $L = \Omega(\frac{\#offlops}{M^{3/2}})$. Assuming that $q \approx m$ the $\#offlops = \frac{m^2(2n-1)}{p}$, then $L = \Omega(\frac{m^2(2n)}{pM^{3/2}})$. Since we know that $\sqrt{M} = mn + \frac{2qr}{p}$, substituting will give us $L = \frac{m^2(2n)}{p\times(mn+\frac{2qr}{p})^3}$. Since for large data sets $q \approx m$, and $n \approx r$, the expression can be approximated as $\frac{2}{mpn^2\times(1+\frac{4}{p}+\frac{12}{p^2}+\frac{8}{p^3})}$. Therefore, $L \approx \Omega(\frac{2}{mpn^2})$.

Theorem 3.5 *The overall runtime lower bound of existing HPC methods is* $\Omega(mn)$ *irrespective of how many processors are used for computations.*

Proof The overall runtime bound can be calculated for existing HPC methods can by summation of L and F, and the communication that is specific to existing algorithms. The summation of these $\frac{qm(2n-1)}{p}+(\frac{2}{mpn^2})+(mn)$ gives us a lower bound on the overall run time which is bounded by $\Omega(mn)$.

Corollary 3.1 *Mass filtering (or other filterings specific to MS data) for candidate generation does not change the communication bounds of* $\Omega(mn)$ *of the current parallel algorithms.*

Proof The communication bounds that we just proved assume that no filtering is taking place for these computations. This ensures that the results are generalizable to all parallel algorithms without getting into method-specific details. However, we show below, that mass-filtering does not change the communication bounds:

Case 1: The mass-filtering takes place on the master node and the database and truncated databases are communicated Worst-case communication bounds are $\Omega(mn)$ since all (or a constant factor) of the database could be communicated at certain nodes and is not in the control of the parallel method since the assumption is that each node will have N/p computations. Assuming that the transmitted parts are a fraction of the number of processors, i.e. q/p, it is easy to see that $\Omega((q/p) * mn)$ computations are needed at the master node, proving that the communication bounds remain unchanged.

Case 2: The mass filtering takes place on each node in parallel. If the mass filtering takes place on each node in parallel, then it needs to communicate $\Omega(mn)$ database to each node and the communication bounds calculated in this paper remain unchanged.

Corollary 3.2 *Fragment-Ion Index scoring similar to MSFragger also does not change the communication bounds of* $\Omega(mn)$ *if the same parallelization method is executed.*

Proof Fragment-Ion index is based on indexing the peaks for each of the theoretical spectra. If the indexing is taking place on the head node then $\Omega(mn)$ communication has to take place to distribute the index on each of the processing nodes.

Theorem 3.6 *We prove that much tighter lower bounds are possible for parallel algorithms (that are yet to be discovered). Combining the lower bounds on W, L, and F will yield lower bounds on the overall run time* ***possible*** *for processor with $M \leq (mn + \frac{2qr}{p})$ memory available. Therefore, the lower bounds* possible *for parallel algorithms is equal to $\Omega(\frac{nmq}{p})$.*

Proof Combining the lower bounds on W, L, and F will yield lower bounds on the overall run time of the existing HPC algorithms. In our theorems, we have proved that $L = (\frac{2}{mpn^2}) + F = \frac{qm(2n-1)}{p} + W = \frac{m^2}{pm+2q}$. This summation gives us a result of $\Omega(\frac{2mnq}{p})$.

Theorem 3.5 shows that the existing HPC algorithm only achieves $\Omega(mn)$ runtime and is independent of the number of processors used. Advantage of using HPC using these algorithms is that a smaller subset (q) of spectra needs to be processed. This advantage, however, vanishes when the size of the theoretical database is increased. Theorem 3.6 predicts $\Omega(\frac{nm^2}{p})$ as the runtime possible when m is approx. equal to q. By approximating q to m results in simple mathematical expressions, it does over-estimate the lower bound of the run time. A more realistic expression could be $\Omega(\frac{nmq}{p})$ which includes both parameters for spectra as well as the size of the database. Nevertheless, an increasing number of PTMs that are used to search for spectra for large-scale non-model systems biology experiments will result in database sizes that are much larger than the number of spectra; therefore being the dominating factor in all communication costs for computational experiments. None of the HPC methods, to date, achieve lower bounds of communication, and computations as dictated by these theorems. Still, significant research efforts are needed to achieve these bounds in theory and practice.

3.5 Meta-Analysis of Results of Current HPC Methods

In order to evaluate how closely the current HPC methods follow the theoretical limits that have been proved in our theorems, we downloaded the results from the following methods [7, 13–18, 21–25]. These methods included MPI-based implementation on distributed-memory machines, Hadoop implementations, and multicore designs. We focused on the results that have been reported for distributed-memory machines but excluded implementations that used Hadoop-based systems since communication patterns and infrastructure information is generally not available from these commercial cloud-based services. CPU-GPU-based implementations were also excluded from this meta-analysis for the same reasons.

We needed to evaluate these HPC methods with metrics that were architecture-specific and system-independent as possible, i.e. cluster-specific advantages are not

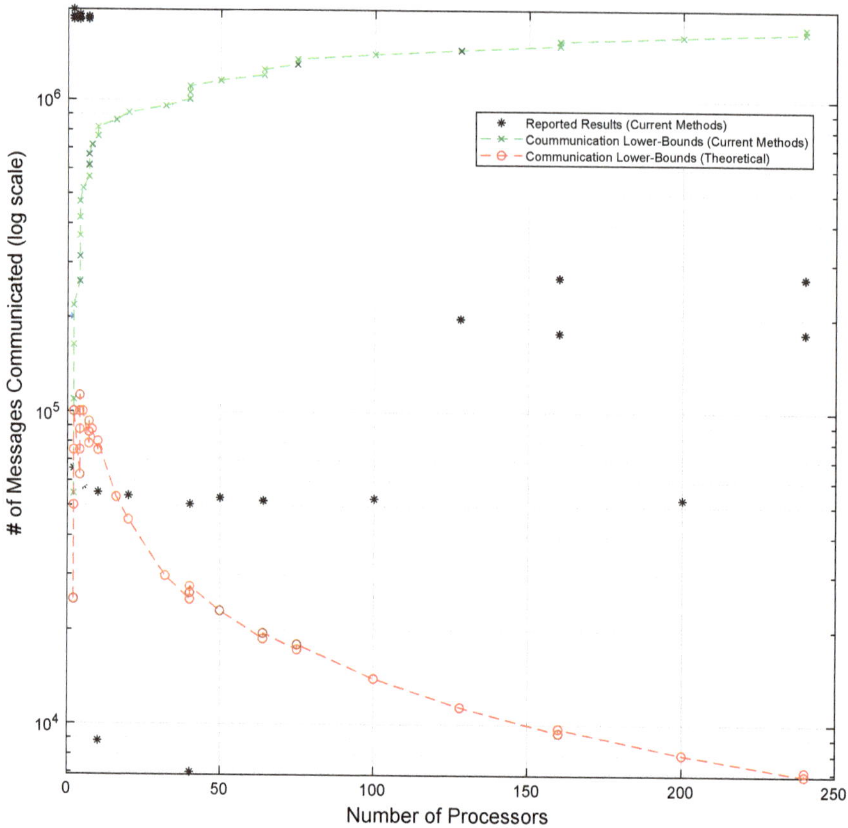

Fig. 3.3 The graph shows the amount of communication that takes place with an increasing number of processors. As can be seen that most of the HPC methods that are listed do not achieve the lower bounds on communication. The gap increases rapidly between the communication required for the state-of-the-art HPC algorithms and the communication that can be theoretically achieved

considered in the evaluation to ensure that it is fair. One of these metrics is the *amount of total communication* that is independent of the machines that might have been used for experimentation and is also in line with our theorems. The other metric that also allows us to evaluate the existing HPC methods is independent of other machines and would ensure fair comparison is *speedups*.

We downloaded all the results [7, 13–18, 21–25] that have been reported till date. The information that we extracted from these methods include database size, number of spectra, reported times, speedups, and memory footprint (not always reported). Using all the information from the published literature, we plotted the communication message that was required to complete the method as a function of theoretical database (spectra length and number are neglected). Both communication bounds that were calculated for the current methods and the theoretical bounds that are possible were plotted. Figure 3.3 shows that the results that are reported are

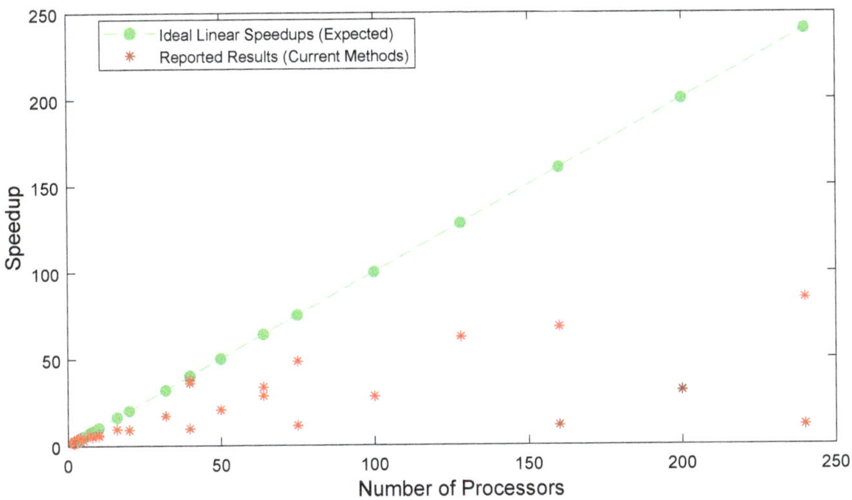

Fig. 3.4 This graph represents the speedups that are reported by the papers and the corresponding linear speedup that can be achieved with an increasing number of processors. Note that the reported results that are listed here are also the results that are depicted in Fig. 3.3 and show a one-to-one correspondence between the amount of communication and the speedups with an increasing number of processors

close to the bounds that we have calculated. As can be seen from the figure the increasing number of processors must need a decreasing number of message (in theoretically communication-optimal algorithm) in contrast to the current real-world implementations. Clearly, not considering communication costs in the design of these parallel algorithms results in much more communication than would be expected from optimal communication-avoiding solutions which is not a norm till date [13].

The second metric that we used was speedups. Speedups are a good metric because the reported results are in fact the speed that one gets compared to the same architecture (with more processors, for instance). This makes speedup an ideal metric that can be used for evaluating the performance of the parallel algorithm without the need to explicitly compare one method with another method (that may be running on a different machine). As expected, in concurrence with our theoretical results, the reported results show decreased speedups with increasing number of processors for all state-of-the-art methods. The results are plotted in Fig. 3.4. As dictated by our theoretical framework, the decrease in the speedup is due to increased communication costs with increasing size of the database and increasing number of processors, i.e. increasing processor does not result in decreased communication due to sharding of the database. Such an optimal communication-avoiding algorithm (which has not been discovered till date) could result in near-linear speedups with increasing number of processors.

3.6 Discussions

Scalable parallel algorithms with provable performance for large-scale MS systems biology studies are essential for making personalized nutrition, microbiome research, and cancer therapeutics a clinical reality. Such scalable algorithms are of more importance when performing systems biology experiments for non-model proteomics, meta-proteomics, and proteogenomics studies where the search-space traversal for peptide(s) with modification can be huge. We have formulated a generalized theoretical framework which allows us to determine the communication bounds that are reached by the existing methods. Our meta-analysis conclusively shows that the performance obtained by existing HPC methods is far less than the theoretically optimal methods and results in sub-optimal speedups with increasing number of processors. Our framework also dictates what are the theoretical limits that can be reached by (yet to be discovered) communication-optimal parallel algorithms. Discovery of such communication-optimal algorithms can result in *provably* superior performance for peptide deduction methods on multicore, GPU, distributed-memory supercomputers, and cloud-computing infrastructure. Such contributions are expected to be significant because they will open up novel and faster ways to analyze MS data for various omics (read: preteomics, proteogenomic, meta-proteomics, etc.) studies considered "too large-scale".

Following are a few caveats that can be used to get a better understanding of the proposed theoretical framework:

1. For the framework, we have assumed a single parallel strategy which is algorithmically the closest one could get for the current HPC methods that have been proposed till date. Variation such as scoring, getting candidate peptides, etc., are not considered part of the parallelization strategy in order to get generalized mathematical bounds. Since data is moved in similar fashion in all existing HPC methods; any inclusion of scoring variations will only result in modification of constant in the proved communication bounds.
2. The theoretical bounds that we have proved are with the assumption that theoretical database and spectra cannot fit inside a single machine memory M. If the size of the data is not that large, then the speedups that one would get could be much higher. However, such a result would be an artifact of the data and or the machine being used, not a generalizable result. This also is in agreement with the meta-analysis which shows that increasing the number of processors does not result in correspondingly linear speedups expected of scalable parallel solutions.
3. We have assumed in our framework that the theoretical database is on the master node which is communicated by the network. One can also assume that the whole database is not communicated (only the FASTA files are communicated from the master processor). However, this assumption will not affect the bounds since the communication costs are now substituted by computations costs, i.e. $O(nm^2/p)$. Therefore, with data distribution method that is used by the current HPC methods will achieve the lower bounds that we have proved. Our meta-analysis of published results also concurs with this argument.

4. Another interesting point to think about is that for calculating our bound we assumed that the whole database is needed for computations. For calculating our bounds, we assume that the whole database is needed for computations. One can argue that only "candidate peptides" are used by real algorithms. However, this reasoning does not affect the communication lower bounds since getting or calculating the candidate spectra still need access to the theoretical spectral database and the communication bottleneck remains unchanged.

3.7 Conclusions

For the past 30 years, efforts have been invested to design efficient peptide deduction algorithms which are, to date, are implemented as numerical algorithms. The HPC methods that have been proposed were designed and implemented as methods that require faster computations with little or no attention to the communication costs of these algorithms. However, the significant increase in the compute power due to Moore's law has made the arithmetic computations faster, and as a result, MS algorithm has shifted from computational arithmetic efficiency to communication bottleneck. However, poor scalability of the existing HPC methods has been considered an artifact of the system or data and has not been investigated.

In this chapter, we have formulated a theoretical framework that can be used to quantify the (communication) performance for the state-of-the-art HPC methods. We presented lower bounds that are reached by these HPC (and serial) methods and lower bounds that could be achieved by parallel algorithms on distributed-memory architectures. To the best of our knowledge, this is the first study to formulate a theoretical framework showing that the existing parallel strategies for MS-based omics data analysis are not achieving the communication bounds that are possible and that continued improvements are needed in this area of research. Meta-analysis of existing HPC methods has also revealed that the theoretical bounds agree with the experimental data that is observed.

Therefore, urgently, parallel algorithms are needed that are communication-optimal, can bridge the gap between theory and practice, and can gracefully scale with an increasing number of processors. In contrast to the present generation of methods, the next generation of HPC methods must be designed by considering *both* computational and communication costs as metrics for efficiency. This next generation of HPC methods can scale (at least) linearly with increasing number of processors, size of the (theoretical) database and would allow massive scale analysis of meta-proteomics, proteogenomics, and non-model omics using MS data and push us one step closer to precision medicine.

References

1. Ballard G, Carson E, Demmel J, Hoemmen M, Knight N, Schwartz O (2014) Communication lower bounds and optimal algorithms for numerical linear algebra. Acta Numer 23:1
2. Ballard G, Demmel J, Holtz O, Schwartz O (2011) Minimizing communication in numerical linear algebra. SIAM J Matrix Anal Appl 32(3):866–901
3. Council NR, et al (2005) Getting up to speed: The future of supercomputing. National Academies Press
4. Demmel J, Eliahu D, Fox A, Kamil S, Lipshitz B, Schwartz O, Spillinger O (2013) Communication-optimal parallel recursive rectangular matrix multiplication. In: 2013 IEEE 27th international symposium on parallel and distributed processing. IEEE, pp 261–272
5. Solomonik E, Bhatele A, Demmel J (2011) Improving communication performance in dense linear algebra via topology aware collectives. In: SC'11: proceedings of 2011 international conference for high performance computing, networking, storage and analysis. IEEE, pp 1–11
6. Eng JK, Fischer B, Grossmann J, MacCoss MJ (2008) A fast SEQUEST cross correlation algorithm. J Proteome Res 7(10):4598–4602
7. Diament BJ, Noble WS (2011) Faster SEQUEST searching for peptide identification from tandem mass spectra. J Proteome Res 10(9):3871–3879
8. Eng JK, McCormack AL, Yates JR (1994) An approach to correlate tandem mass spectral data of peptides with amino acid sequences in a protein database. J Am Soc Mass Spectrom 5(11):976–989
9. McIlwain S, Tamura K, Kertesz-Farkas A, Grant CE, Diament B, Frewen B, Howbert JJ, Hoopmann MR, Kall L, Eng JK et al (2014) Crux: rapid open source protein tandem mass spectrometry analysis. J Proteome Res 13(10):4488–4491
10. Kong AT, Leprevost FV, Avtonomov DM, Mellacheruvu D, Nesvizhskii AI (2017) Msfragger: ultrafast and comprehensive peptide identification in mass spectrometry-based proteomics. Nat methods 14(5):513
11. Yates III JR (2019) Proteomics of communities: metaproteomics
12. Heck M, Neely BA (2020) Proteomics in non-model organisms: a new analytical frontier. J Proteome Res
13. Kulkarni G, Kalyanaraman A, Cannon WR, Baxter D (2009) A scalable parallel approach for peptide identification from large-scale mass spectrometry data. In: 2009 international conference on parallel processing workshops. IEEE, pp 423–430
14. Li C, Li K, Li K, Lin F (2019) MCtandem: an efficient tool for large-scale peptide identification on many integrated core (MIC) architecture. BMC Bioinform 20(1):397
15. Sun J, Chen B, Wu F-X (2014) An improved peptide-spectral matching algorithm through distributed search over multiple cores and multiple CPUS. Proteome Sci 12(1):18
16. Duncan DT, Craig R, Link AJ (2005) Parallel tandem: a program for parallel processing of tandem mass spectra using PVM or MPI and x! tandem. J Proteome Res 4(5):1842–1847
17. Bjornson RD, Carriero NJ, Colangelo C, Shifman M, Cheung K-H, Miller PL, Williams K (2008) X!! tandem, an improved method for running x! tandem in parallel on collections of commodity computers. J Proteome Res 7(1):293–299
18. Li C, Li K, Chen T, Zhu Y, He Q (2019) SW-tandem: a highly efficient tool for large-scale peptide sequencing with parallel spectrum dot product on Sunway TaihuLight. Bioinformatics (Oxford, England) 35(19):3861–3863
19. Ballard G, Demmel J, Holtz O, Lipshitz B, Schwartz O (2012) Communication-optimal parallel algorithm for strassen's matrix multiplication. In: Proceedings of the twenty-fourth annual ACM symposium on parallelism in algorithms and architectures, pp 193–204
20. Haseeb M, Saeed F (2021) High performance computing framework for tera-scale database search of mass spectrometry data. Nat Comput Sci 1(8):550–561
21. Li C, Li K, Li K, Xie X, Lin F (2019) Swpepnovo: an efficient de novo peptide sequencing tool for large-scale MS/MS spectra analysis. Int J Biol Sci 15(9):1787

22. Baumgardner LA, Shanmugam AK, Lam H, Eng JK, Martin DB (2011) Fast parallel tandem mass spectral library searching using GPU hardware acceleration. J Proteome Res 10(6):2882–2888
23. Pratt B, Howbert JJ, Tasman NI, Nilsson EJ (2012) MR-tandem: parallel x! tandem using hadoop MapReduce on amazon web services. Bioinformatics 28(1):136–137
24. Li C, Chen T, He Q, Zhu Y, Li K (2016) Mruninovo: an efficient tool for de novo peptide sequencing utilizing the hadoop distributed computing framework. Bioinformatics 33(6):944–946. https://academic.oup.com/bioinformatics/article-pdf/33/6/944/25147928/btw721.pdf,https://doi.org/10.1093/bioinformatics/btw721https://doi.org/10.1093/bioinformatics/btw721
25. Kalyanaraman A, Cannon WR, Latt B, Baxter DJ (2011) Mapreduce implementation of a hybrid spectral library-database search method for large-scale peptide identification. Bioinformatics 27(21):3072–3073. https://academic.oup.com/bioinformatics/article-pdf/27/21/3072/16901315/btr523.pdf, https://doi.org/10.1093/bioinformatics/btr523

Chapter 4
High-Performance Computing Strategy Using Distributed-Memory Supercomputers

Fahad Saeed and Muhammad Haseeb

4.1 Introduction

Database peptide search is the most commonly employed computational technique to deduce peptides from the experimentally obtained mass spectrometry data [2]. In this technique, the experimental spectral data are compared against a protein sequence database through various search algorithms in order to assign the correct peptide sequence to each experimental spectrum [3]. Since the experimental spectra data (histogram-like data) and the peptide sequence data (text data) are not one-to-one comparable, database search simulates the mass spectrometry process in silico to generate theoretical spectra from the peptide sequences in the database [3]. Several other techniques also exist for complete or partial inter-conversion between the two data domains including sequence-tagging [4, 5], de novo [6–8], and their combinations [9, 10]. In general, the workflow of any modern database peptide search algorithm involves the following steps [3]:

1. Clean, denoise, reconstruct, and pre-process the experimental data.
2. Simulate in silico mass spectrometry to generate a database of theoretical spectra.
3. Optional: Index the converted database for faster search.
4. Search the processed experimental data against the processed (and indexed) database.
5. Post-processed the results, compute statistical significance of the matches, compute false discovery, and so on.

Many software frameworks have been proposed in the past three decades to speed up the database search workflow, most of which have focused on reducing the number of required computations through data cleaning, aggressive database filtration, and de novo-assisted approaches [4, 5, 9–21]. Therefore, these advanced algorithms now increasingly require both CPU and memory resources for optimal performance.

Some parts of this chapter may have appeared in [1].

F. Saeed and M. Haseeb, *High-Performance Algorithms for Mass Spectrometry-Based Omics*, Computational Biology 34,
https://doi.org/10.1007/978-3-031-01960-9_4

Consequently, the number of computations performed per memory transaction (arithmetic intensity) [22] now upper bounds the performance of most modern database search algorithms. This situation is further aggravated as the database index incorporates post-translational modifications (PTMs) leading to several hundred giga- to terabyte-scale databases [11, 23]. As shown in several big data fields [24], memory contention upper bounds can be alleviated by effectively dividing-and-conquering the computational problem across a distributed-memory supercomputer. Therefore, in this chapter, we focus on an efficient distributed-memory computational design for accelerating the database peptide search problem.

4.1.1 Background

Many computational frameworks such as X!!Tandem [25, 26], MR-Tandem [27], SW-Tandem [28], MCtandem [29], Bolt [30], MS-PyCloud [31], and UltraQuant strive to accelerate the database peptide search on supercomputers. However, the major limitation in all existing frameworks is the inefficient use of the distributed architecture, leading to the same memory contention that limits the shared-memory algorithms. This inefficiency stems from their parallel design (divide-and-conquer strategy) that involves partitioning the experimental data across parallel nodes, each having a replica of the complete (gigantic) database index. For instance, ParallelTandem [25] runs parallel instance of X!Tandem using MPI or PVM. X!!Tandem [26] implements an internally parallelized X!Tandem using MPI. MR-Tandem [27] parallelize through Map-Reduce model allowing for better load balance. MCtandem [29] and SW-Tandem [28] implement accelerator-offloaded versions of the X!Tandem with some extra optimizations. Furthermore, Bolt [30] implements a parallel MS-Fragger, MS-PyCloud [31] parallelizes MS-GF+ [32] in cloud-computing setting and UltraQuant parallelizes MaxQuant37 using the same straightforward parallel design.

In 2009, Kulkarni et al. [33] proposed two preliminary parallel designs that correctly leverage the distributed memory by splitting the large database index among parallel nodes reducing memory contention per node. However, the data stream-based search workflow employed by these parallel models limited the parallel efficiency to 50% due to frequent inter-node communications and on-the-fly computations performed to merge local results into collective results at the end.

4.1.2 Problem Statement

Identification of peptides from mass spectrometry data is a fundamental step in computational proteomics. Existing parallel database search algorithms are based on inherently shared-memory designs that do not optimally scale on distributed-memory architectures. This leaves a gap for a parallel strategy capable of deriving maximum performance from modern high-performance computing (HPC) systems. We introduce HiCOPS [1], a search algorithm-independent parallelization strategy for the acceleration of generic database peptide search workflows onto distributed-memory supercomputers, achieving impressive computational speeds and near-optimal parallel scalability.

4.2 The HiCOPS Framework

HiCOPS's design is based on the single program, multiple data (SPMD) Bulk Synchronous Parallel (BSP) [34] computational model where the computations are performed by a pool of independent asynchronous parallel computing nodes. The nodes execute the required algorithmic computations in *supersteps* [34] and synchronize after each superstep, if required. A superstep is defined as a series or a set of asynchronously executable algorithmic work. HiCOPS splits the database peptide search workflow into the following four supersteps as also shown in Fig. 4.1.

1. Split the theoretical database across the compute nodes in a load-balanced fashion and create local database indices.
2. Partition the experimental MS/MS data set across the compute nodes and locally pre-process, index, denoise, and re-construct the data as needed.
3. On each parallel node, search the entire experimental MS/MS data set (in batches) against the local database index using the provided search algorithm.
4. Assemble a partition of the intermediate results produced in the last step at each parallel node and communicate back the final results to their origin nodes.

4.2.1 Database Indexing

The peptide sequences along with their post-translationally modified sequences are partitioned across the parallel nodes in a load-balanced fashion using the LBE [35] algorithm. The LBE algorithm is supplemented with a new metric, called *Mod Distance* (Δm) [1], to allow separation of database entries (peptides or variants) that have the same *Edit Distance* (Δe) [35]. The extracted partition of the database is then locally indexed at all parallel nodes using the supplied indexing algorithm.

Scalability: This superstep generates similar sized database slices in $O(D)$ time followed by local indexing in $(D/p \ \log \ D/p)$ time, where D is the size of complete search space and p is the number of parallel nodes. Therefore, the scalability of this step is limited by the partitioning step.

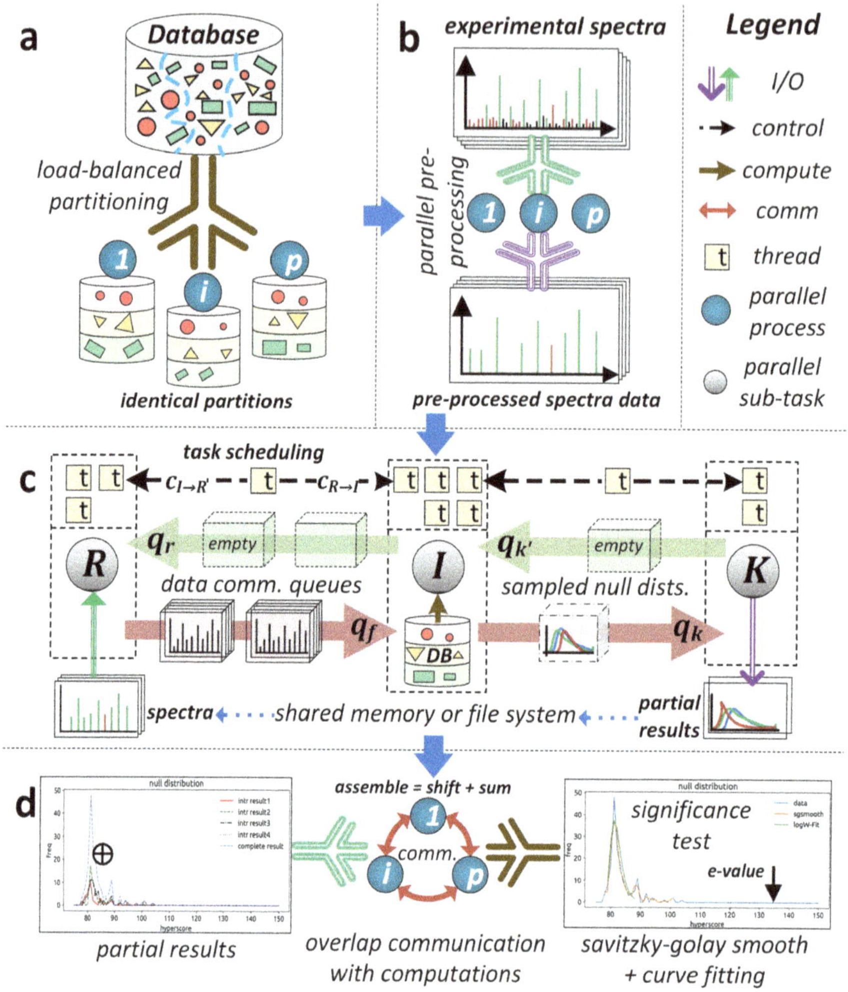

Fig. 4.1 **a Superstep 1**: The massive search space of theoretical spectra (shown as shapes) is partitioned among parallel MPI processes in a load-balanced (clustering (stripes) + scattering) fashion **b Superstep 2**: The experimental MS/MS spectra data pre-processed in data-parallel fashion. **c Superstep 3**: Per-process view of the parallel database peptide search workflow executed in hybrid task and data-parallel fashion at all HiCOPS tasks. On each process, three parallel tasks R, I, and K work in a producer-consumer pipeline to load the experimental data search it against the local search space, and communicate the local results to the shared memory, respectively. The thread allocation to each task is managed through a task scheduling algorithm to maintain pipeline synergy. **d Superstep 4**: The local results from the last superstep are assembled into complete results, which are inter-communicated between the nodes for the final output

4.2.2 Experimental Data Pre-processing

This superstep is executed in an *embarrassingly parallel* fashion by randomly partitioning the experimental MS/MS spectra data across parallel nodes followed by the execution of supplied pre-processing algorithms. The pre-processed data is written back to the shared file system on completion for subsequent use. Note that this step may be skipped in subsequent runs altogether if the pre-processed data is available.

Scalability: This superstep exhibits a coarse-grained embarrassingly parallel profile as equal-sized chunks of experimental data can be processed by parallel nodes in near-equal amount of time. In real-world scenario, this superstep may incur some overheads due to load imbalance and data communication overheads.

4.2.3 Parallel Database Peptide Search

This is the most important superstep in HiCOPS and executes about 80–90% of the overall algorithmic word. In this superstep, the parallel nodes search the pre-processed experimental spectra against their local database index. This superstep is designed through a hybrid (task and data-parallel) producer-consumer pipeline model explained as follows. Three parallel tasks, namely R, I, and K are forked that load batches of pre-processed experimental data into the main memory, execute the database search, and write intermediate results to the shared file system, respectively. The number of parallel cores assigned to each task at a given time is governed by a forecasting-based task-scheduling algorithm explained in later sections. The data flow between the pipeline is designed using memory buffers and queues to handle speed mismatches between producers and consumers.

Scalability: This superstep executes a complex pattern of compute, memory, and communication operations due to the involved data flows and random database accesses required. Therefore, to derive the maximum performance from the system, we introduce several optimizations including task scheduling, data buffering, and sampling (discussed in Sect. 4.3). These optimizations ensure that the system resources are never under- or over-utilized to avoid overheads. The task R reads the pre-processed experimental data from the storage in linear time, the task I searches the data against the local database in $O(sND)$ time, where s is the complexity of the supplied search algorithm, N is the size of experimental data set, and D is the size of local database. Finally, the task K communicates the local results in $O(r)$ time, where r is the size of intermediate results data. Since this superstep executes in a hybrid fashion, the overall time for this step can be written as $O(sND) + \theta$ where $O(sND)$ is time for task I and θ is the extra time needed by K to finish after I is done. The above equation is true since the task R (producer) must complete before I (consumer) which is the main step plus some extra time to complete the pipeline given by θ.

4.2.4 Assembling the Local Results

This superstep assembles the intermediate results produced in the last superstep in a hybrid parallel fashion. Two parallel tasks W and L are spawned, where W performs the assembly and L communicates the assembled global results to the other parallel HiCOPS nodes. W assembles the intermediate results into final results through accumulation and digital signal shift operations. The complete null distribution is constructed and statistical significance (expected values) of top-scoring hits is computed either using Linear Tail-Fit method or (log-Weibull) curve fitting method. Figure 4.1d illustrates the score assembly process. Static thread scheduling is used in this superstep and all available cores (c_{p_i}) are assigned to W with 1 over-subscribed thread assigned to L. Since an all-to-all data exchange may be needed in worst-case scenario, the number of inter-process communication operations are upper bound bounded by $O(P^2)$. Finally, once the two tasks W and L complete (join), the final results are written to the file system in data-parallel fashion.

Scalability: W performs the assembly in time: $O(xrp)$, where r is the number of results, p is the number of HiCOPS processes, and x is the runtime of the curve fitting algorithm (typically linear or quadratic regression).

4.3 Optimizations

4.3.1 Task Scheduling

Recall that the data flow between the pipeline tasks in the database search superstep is designed using buffer queues. The status of the queues at any point during the execution also indicates the relative speed of the three tasks. For example, the number of experimental data batches in queue: $R \rightarrow I$ at any time indicates their relative speeds. If the queue is getting fuller, it means that the data production is faster than the consumption speed and vice versa. There are certain edge cases as well such as the initial state of the queues and the time to produce at least one batch of data which are handled by keeping a record of the change and accumulation of the production to consumption delay by the scheduling algorithm. Three regions are defined for the queue $R \rightarrow I$, i.e. $r1$ = (data < 5), $r2$ = ($5 \leq$ data < 15), and $r3$ = (data $\geq$ 15) which alert the scheduling algorithm if a change may be needed in the near future. The task scheduling algorithm dynamically assigns threads to each of the tasks R, I, and K using the Algorithm 2.

Algorithm 2: Task Scheduling in Superstep 3

Data: buffer queue: $R \rightarrow I$ (q_F), per batch halt time (t_h), thresholds for minimum halt (t_{min}), surge (t_{surge}), and max accumulation (t_{acc})

Result: Optimal thread allocation between subtasks R, I, and K

/* Local variable to record halt time accumulation */

$t_{cumu} \leftarrow 0$

/* While data is being produced */

while ($b_{rem} > 0$) **do**

    /* If halt time is larger than t_{min} */

    if $t_{halt} \geq t_{min}$ **then**

        $t_{cumu} += t_{halt}$

        if $t_{cumu} > t_{acc}$ *or* $t_{halt} > t_{surge}$ **then**

            $move\_thread(src = I, dst = R)$ $t_{cumu} \leftarrow 0$

    else

        if $q_F.size() < 5$ *and* $| r | == 0$ **then**

            $move\_thread(src = I, dst = R)$ $t_{cumu} \leftarrow 0$

        else if $q_F.size() > 15$ *and* $| r | > 1$ **then**

            $move\_thread(src = R, dst = I)$

return $move\_Allthreads(src = R, dst = I)$

4.3.2 Communication Optimization

Each parallel node executing HiCOPS produces intermediate results for each experimental spectrum consisting: the top-scoring search hit (8 bytes) and the local null distribution of the scores (2048 bytes) totaling a data packet = ~2 KB per experimental spectrum. Since most data sets on mass spectrometry data repositories contain several hundred thousand to millions of spectra, communicating $2KB \times p^2$ data over the network would result in loss in performance. To mitigate this, we implement two methods that exploit the produced null distribution information which is known to follow a log-Weibull curve. The first method uses a regression technique to fit a standard log-Weibull curve into the null distribution keeping only the curve parameters: μ and β (8 bytes) and discarding the whole 2KB distribution data. The advantage of this technique is that the payload size reduces from 2KB to 16bytes. The disadvantages, however, include loss in the accuracy of results when assembled as well as the extra computations required for regression. The extra computations overhead could be reduced by offloading the task to GPUs, FPGAs, or other accelerators if available. The second method directly samples the null distribution by first locating the mean of the curve and then extracting at most s ($s = 120$ default) intense samples around the mean. Further, the samples along the head of the curve are prioritized since the tail in log-Weibull distributions are usually skewed (flat) and could be easily interpolated if needed. This allows us to fit the local results per experimental spectrum within 256 bytes (in contrast to 2KB) significantly reducing the communication overheads. The sampling algorithm is depicted in Algorithm 3.

Algorithm 3: Sampling Algorithm

Data: local null distribution (d_l), number of samples to extract (s_{max})
Result: Sampled null distribution (d_s)
/* find sampling space along x-axis */
$x1 = findMin(d_l[x])$
$x2 = findMax(d_l[x])$
/* find the distribution's μ value */
$\mu = AverageMeanAndMedian(d_l, x1, x2)$
/* pick samples until s_{max} */
/* check if we have more than s_{max} samples to begin with */
if $x2 - x1 > s_{max}$ **then**
  /* sample until $s = s_{max}$ */
  while $s < s_{max}$ **do**
    /* sample around μ by prioritizing curve head over the skewed-tail */
    $sa = samplebyPriority(\mu, x1, x2)$
    /* append to d_s */
    $d_s.append(sa)$;
    /* optimize for maximum distribution area picked */
    $optimize(area(d_s), area(d_l))$
else
  /* simply return the unsampled d_l */
  $d_s = d_l$
return dl

4.4 Results

UniProt *Homo sapiens* and UniProt *SwissProt* databases were used with combinations of commonly occurring post-translational modifications (PTMs) to emulate increasingly larger theoretical-spectra databases. Furthermore, combinations of several PRIDE Archive data sets (Accession numbers: PXD: 009072, 020590, 015890, 007871, 010023, 012463, 013074, 013332, 014802, and 015391) were used as data sets across our experiments. All experiments were run on the Extreme Science and Engineering Discovery Environment (XSEDE) Supercomputers Comet cluster located at San Diego Supercomputing Center (SDSC). Note that the minimum number of parallel nodes required by HiCOPS in an experiment must be $\geq D/M$, where D is the size of database index and M is the main memory per node.

4.4.1 Experimental Settings

The databases were digested in silico using Trypsin as an enzyme (fully tryptic) with two allowed missed cleavages, peptide lengths between 6 and 46, and peptide masses between 500 and 5000Da using OpenMS toolkit. The theoretical spectra were simulated by generating b- and y-ions up to +3 charge with zero isotope error and no decoys. Cysteine carbamidomethylation was set as a fixed modification. The maximum number of allowed modified residues (amino acid letters) per peptide was set to 5. Restricted search peptide precursor filter was set to ±5Da to cover for differences in mass calculation errors (average or monoisotopic masses) and open-search peptide precursor filter was set to $\geq \pm 100$Da. We will describe each performed experiment in terms of the database and data set sizes and the filters used as a tuple $t = (D, q, \delta M, \delta F)$, where D is the theoretical spectra database size in 100 million spectra, q is the experimental data set size in 1 million spectra, δM is the peptide precursor mass filter in ±1Da, and δF is the fragment-ion mass filter in ±1Da. Note that if δF (fourth element) is not specified in a tuple, it means that it is set to ±0.01Da.

The experimental MS/MS spectra pre-processing was set to minimum for fairness across search tools. These settings were as follows: allowed precursor mass and charge range to 500–5000Da and +1 to +4, respectively, min. peaks matched for PSM candidacy: 4, min. database hits for statistical modeling: 4, denoising: peak-picking top 100 (by intensity), all other processing: off.

4.4.2 Correctness Analysis

We broke down our correctness analysis into two parts, i.e. correctness of the parallel design and correctness of peptide identification. For the first part, we compared the HiCOPS's results by running it on an increasing number of parallel nodes against the ones produced by serial HiCOPS execution. The results in Figs. 4.2 and 4.3 show over 99.6% consistency between serial and parallel runs up to 3 decimal points for several experiments with varying settings. The small number of inaccurate results stemmed from precision losses in sampling and assembly steps. For the second part, we compared the peptide identifications from HiCOPS and MSFragger. For this, we implemented an MSFragger-like fragment-ion index-based shared-peak counting coupled hyperscore search algorithm within HiCOPS. In our setup, we performed three sets of experiments in both restricted- and open-search settings. These experiments in the tuple form are given as: $c_1 = (0.18, 0.86, 0.01)$, $c_2 = (0.66, 1.51, 0.01)$, $c_3 = (0.80, 1.51, 0.01)$, $c_4 = (0.18, 0.86, 2)$, $c_5 = (0.66, 1.51, 1)$, $c_6 = (0.80, 1.51, 2)$. The obtained results from both tools depicted a close correlation between hyperscores computed by both tools in restricted-search mode (pearson coefficient $R \geq 90\%$), which dropped to about $R \geq 75\%$ in open-search setting. There are many reason for this drop including the internal data processing, search and re-ranking algorithms employed MSFragger. Since MSFragger is not an open-source tool, the exact reason could not be fully determined. Remember that the peptide identifications produced by HiCOPS depend on the supplied algorithms and not the parallel design itself.

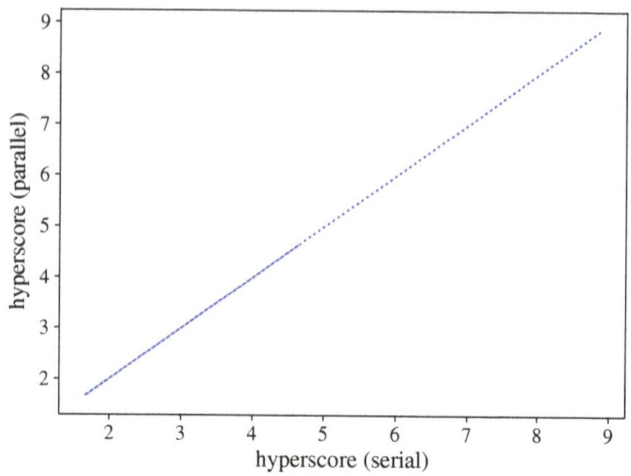

Fig. 4.2 HiCOPS computed hyperscores in serial and parallel runs relate by $y = x + \epsilon$

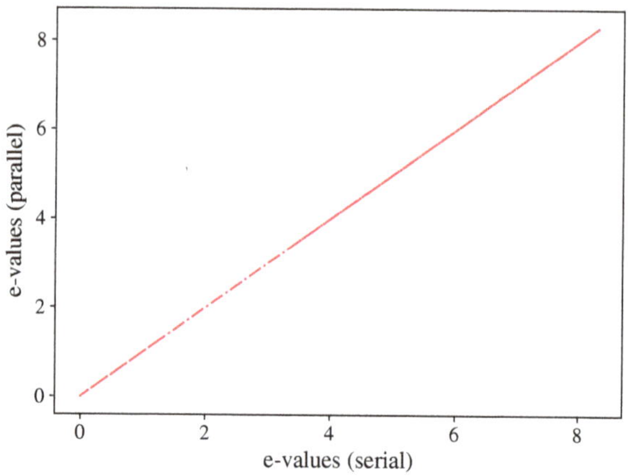

Fig. 4.3 HiCOPS computed e-values in serial and parallel runs relate by $y = x + \epsilon$

4.4.3 Speed Comparison

We compared the HiCOPS-based hyperscore algorithm against various database peptide search tools in both serial and parallel mode across six experiments of increasing algorithmic workload sizes given in tuple form as s_1 = (0.008, 0.935, 0.1), s_2 = (0.008, 0.935, 5), s_3 = (0.07, 1.51, 5), s_4 = (0.935, 1.515, 0.1), s_5 = (0.935, 1.515, 5), s_6 = (2.13, 3.89, 1). The parallel versions of the serial database search engines, if not available, were implemented by spawning multiple instances of their binary across a cluster, each searching a partition of the experimental MS/MS data set. We used our implemented shared-peak coupled hyperscore search algorithm [11] to compare HiCOPS's performance with other tools. The obtained results depict that the HiCOPS-based search leverages its optimized parallel design to provide, on average, about 10× speedup over several other tools in all parallel configurations (See Tables 4.1, 4.2, 4.3, 4.4, 4.5, 4.6). In order to alleviate the effect of search algorithm-based speedups across tools, we compared HiCOPS with MSFragger

Table 4.1 Runtime for experiment: $s_1 = (0.008, 0.935, 0.1)$ in seconds for several tools with increasing number of parallel nodes

Tool	n = 1	n = 2	n = 4	n = 8	n = 16
HiCOPS	221.12	166.321	126.358	113.538	134.869
MSFragger	299.4				
SW-Tandem	1015	992	1002	999	1019
Crux	2470				
X!!Tandem	4980	2445	1279	690	360

Table 4.2 Runtime for experiment: $s_2 = (0.008, 0.935, 5)$ in seconds for several tools for increasing number of parallel nodes (n)

Tool	n = 1	n = 2	n = 4	n = 8	n = 16	n = 32	n = 64
HiCOPS	349.52	188.1	135.792	115.186	101.796	101.089	143.941
MSFragger	521.2818						
X!Tandem	18645						
X!!Tandem	115000	57700	29050	14600	7400	3720	1980
SW-Tandem	17979	17094	15366	14344	15124	15097	14973

Table 4.3 Runtime for experiment: $s_3 = (0.07, 1.51, 5)$ in seconds for several tools with increasing number of parallel nodes (n)

Tool	n = 1	n = 2	n = 4	n = 8	n = 16	n = 32
HiCOPS	1850.135	960.589	520.743	315.696	198.305	134.474
MSFragger	12037.82	6692.2	3599.6	2151.46	984.72	987
Comet	44700	33240	19950	11482	5669.1	3601.05
X!!Tandem				29947.28	15612.64	8228.57
MSGF+ (1Da)					46800	25200

(similar algorithm) in more detail. The results show that HiCOPS depicts much better overhead performance (lower I/O and load imbalance overheads), even though the compute time for both tools was similar (See Tables 4.3, 4.4, 4.5, 4.6 for runtimes and Tables 4.7, 4.8, 4.9, 4.10 for corresponding percentage overhead times).

4.4.4 Performance Evaluation

HiCOPS's parallel performance was analyzed using 12 experiments of varying sizes (i.e. search space, data set and open-/closed-search settings). These experiments in their tuple form are given as $exp_1 = (0.3, 0.84, 0.1)$, $exp_2 = (0.3, 0.84, 2)$, $exp_3 = (3.89,$

Table 4.4 Runtime for experiment: $s_4 = (0.935, 1.515, 0.1)$ in seconds for several tools with increasing number of parallel nodes (n)

Tool	n = 1	n = 2	n = 4	n = 8	n = 16
HiCOPS	965.56	557.549	371.585	262.16	213.622
MSFragger	39439	20043	8591.54	4750.7	3389.6
X!Tandem	1710000				
Crux	875500				

Table 4.5 Runtime for experiment: $s_5 = (0.935, 1.515, 5)$ in seconds for several tools with increasing number of parallel nodes (n)

Tool	n = 1	n = 2	n = 4	n = 8	n = 16	n = 32	n = 64
HiCOPS	89682.5	22541.98	6607.923	2797.644	1445.38	807.054	485.541
MSFragger	203954.7	83900	53960	23500	16800	11890	6030

Table 4.6 Runtime for experiment: $s_6 = (2.13, 3.89, 1)$ in seconds for several tools with increasing number of parallel nodes (n)

Tool	n = 4	n = 8	n = 16	n = 32	n = 64
HiCOPS	15832.72	3086.035	1405.969	858.5937	596.3815
MSFragger		19295.5	9952.23	5973.42	3380.52
Comet (10Da)			30747.6	20796.6	12627.6

Table 4.7 Percentage MSFragger runtime spent as I/O overhead in speed comparison experiments (Tables 4.3, 4.4, 4.5, 4.6) for increasing number of parallel nodes (n)

Exp. No.	n = 1	n = 2	n = 4	n = 8	n = 16	n = 32	n = 64
3	54.73001	36.77117	37.94588	37.57913	42.5197	32.70517	
4	89.41652	83.91858	81.0623	42.12432	19.17925		
5	23.595	25.43838	15.22387	7.24766	3.869643	2.24222	2.824212
6				63.00951	57.55996	29.43372	14.72555

Table 4.8 Percentage HiCOPS runtime spent as I/O overhead in speed comparison experiments (Tables 4.3, 4.4, 4.5, 4.6) for increasing number of parallel nodes (n)

Exp. No.	n = 1	n = 2	n = 4	n = 8	n = 16	n = 32	n = 64
3	0.831183	1.611824	4.65796	4.973139	10.05522	10.22056	
4	1.56	3.0322	10.82498	7.657919	9.497617		
5	0	0.071812	0.608724	0.717604	1.403714	7.378193	8.38405
6			2.559578	3.584081	7.430818	12.48344	18.53344

Table 4.9 Percentage MSFragger runtime spent as load imbalance overhead in speed comparison experiments (Tables 4.3, 4.4, 4.5, 4.6) with increasing number of parallel nodes (n)

Exp. No.	n = 1	n = 2	n = 4	n = 8	n = 16	n = 32	n = 64
3	0	1.02	6.99	20.47	37.66	33.59	
4	0	2.09	43.79	35.47	32.64		
5	0	6.04	12.3	9.49	61.27	137.04	108.62
6				28.65	34.48	69.037	39.48

Table 4.10 Percentage HiCOPS runtime spent as load imbalance overhead in speed comparison experiments (Tables 4.3, 4.4, 4.5, 4.6) with increasing number of parallel nodes (n)

Exp. No.	n = 1	n = 2	n = 4	n = 8	n = 16	n = 32	n = 64
3	0	0.204832	0.152784	1.74169	5.498741	8.084369	8.401276
4	0	0.214879	7.623076	2.620038	4.615141		
5	0	0.389582	2.180723	4.12997	4.386946	7.95819	5.386342
6				2.203713	2.581644	0.979308	3.581092

0.07, 5), exp_4 = (1.51, 2.13, 5), exp_5 = (6.1, 0.93, 5), exp_6 = (3.89, 7.66, 5), exp_7 = (1.51, 19.54, 5), exp_8 =(1.6, 38.89, 5), exp_9 = (3.89, 15.85, 5), exp_{10} = (3.89, 1.08, 5), exp_{11} = (1.58, 2.13, 1), and exp_{12} = (0.305, 0.847, 5). Recall that the minimum number of parallel nodes required by HiCOPS an experiment must be $\geq D/M$, where D is the size of database index and M is main memory per node. Therefore, the speedup and strong-scale efficiency calculations in performance analysis were done *relative* to the base case (smallest experiment with nodes > $\geq D/M$). The results show between 70 and 80% *relative* strong-scale efficiency for HiCOPS on average for sufficiently large experiments (See Tables 4.11, 4.12, 4.13). Experiments with smaller workloads deteriorate in parallel efficiency according to the Amdahl's law. Hyper-linear efficiency was also observed for larger experiments due to sharp decrease in memory contention with increase in number of parallel nodes. To verify this, we also instrumented hardware counters that depict an improved instructions per cycle and cache misses performance per node leading to lower memory contention in this scenario (See Tables 4.14, 4.15, 4.16).

4.4.4.1 Overhead Evaluation

We further quantified the overheads in HiCOPS by measuring several metrics including load imbalance, inter-node communication, pipeline stalls, I/O, and so on in the above 12 experiments. The results depict that the I/O and load imbalance constitutes about 10% of the runtime each and the inter-node communications constitute about 5% of the overall runtime keeping the overall overhead timeless than 25% (See Tables 4.17, 4.18, 4.19, 4.20).

Table 4.11 HiCOPS's experimental runtime in seconds for the 12 experiments (exp_i) for increasing number of parallel nodes (n)

Exp. No.	n = 1	n = 2	n = 4	n = 8	n = 16	n = 32	n = 64	n = 72
1		136.1	96.1	84.86	76.1			
2		402.7	171.8	108.93	91.31			
3	1850.14	960.59	520.74	315.7	198.31	134.47		
4			5832.72	3086.04	1405.97	858.59	596.38	
5		40168.01	12887.29	5516.6	3024.14	1785.71	1137.6	
6					2924.2	1395.58	873.23	
7						6911.85	3205.26	
8							1755.31	1634.56
9						4554.66	2418.61	
10		6721.3	2686.52	1301.51	728.86	424.01	296.96	
11			1397.34	751.81	486.53	344.5	297.1	
12		888.64	329.21	155.91	110.14			

Table 4.12 HiCOPS's *relative speedup* for increasing number of parallel nodes (n)

Exp. No.	n = 1	n = 2	n = 4	n = 8	n = 16	n = 32	n = 64	n = 72
1		1	1.42	1.6	1.79			
2		1	2.34	3.7	4.41			
3	1	1.93	3.55	5.86	9.33	13.76		
4			1	1.89	4.15	6.79	9.78	
5		1	3.12	7.28	13.28	22.49	35.31	
6					1	2.1	3.35	
7						1	2.16	
8							1	1.07
9						1	1.88	
10		1	2.5	5.16	9.22	15.85	22.63	
11			1	1.86	2.87	4.06	4.7	
12		1	2.7	5.7	8.07			

4.5 Discussion

High-performance computing (HPC) has shifted towards heterogeneous computing [36] as the Graphical Processing Units (GPUs) become integral components of the modern top 500 supercomputers [37]. Therefore, our future efforts focus on completely revamping the currently proposed SPMD-BSP-based HiCOPS parallel design to fully leverage GPUs and other hardware accelerators in conjunction with symmetric multiprocessing nodes to further accelerate the database search. As shown by our experimental results, HiCOPS's peptide identification rates are limited by the (sup-

Table 4.13 HiCOPS's *relative* strong-scale efficiency for the 12 experiments (exp_i) for increasing number of parallel nodes (n)

Exp. No.	n = 1	n = 2	n = 4	n = 8	n = 16	n = 32	n = 64	n = 72
1		1	0.71	0.4	0.22375			
2		1	1.17	0.925	0.55125			
3	1	0.965	0.8875	0.7325	0.583125	0.43		
4			1	0.945	1.0375	0.84875	0.61125	
5		1	1.56	1.82	1.66	1.405625	1.103438	
6					1	1.05	0.8375	
7						1	1.08	
8							1	0.951
9						1	0.94	
10		1	1.25	1.29	1.1525	0.990625	0.707188	
11			1	0.93	0.7175	0.5075	0.29375	
12		1	1.35	1.425	1.00875			

Table 4.14 HiCOPS's instructions per cycle (ipc) for the 12 experiments (exp_i) for increasing number of parallel nodes (n)

Exp. No.	n = 1	n = 2	n = 4	n = 8	n = 16	n = 32	n = 64
1		0.658605	0.62594	0.606399	0.598779		
2		0.761496	1.194591	1.28988	1.265229		
3	1.172292	1.184356	1.203304	1.193149	1.161948	1.126351	
4			0.936649	0.919458	0.919458	0.919458	0.919458
5		0.56322	0.975077	1.163958	1.176262	1.18136	1.170231
6					0.969719	1.19811	1.274765
9						0.97069	1.204607
10		0.85263	1.131857	1.249126	1.257584	1.24926	1.214691
11			0.936649	0.919458	0.913205	0.905938	0.932812
12		0.627035	1.102769	1.410214	1.40052		

plied) executed algorithms for data pre-processing, searching, and post-processing. Therefore, to improve on this, we also focus our future efforts towards the in-house development of existing and new algorithms, and machine- and deep learning models [38, 39] for various HiCOPS's supersteps.

Table 4.15 HiCOPS's LLC cache misses/total cache misses (lpc) for the 12 experiments (exp_i) for increasing number of parallel nodes (n)

Exp. No.	n = 1	n = 2	n = 4	n = 8	n = 16	n = 32	n = 64
1		0.357922	0.356453	0.339776	0.318339		
2		0.227645	0.097253	0.069666	0.078104		
3	0.022755	0.023158	0.02732	0.029027	0.025319	0.02616	
4			0.104159	0.10849	0.10849	0.10849	0.10849
5		0.22868	0.08208	0.0218	0.020076	0.023515	0.028213
6					0.109903	0.05892	0.040184
9						0.08764	0.035651
10		0.143212	0.062753	0.0326	0.033188	0.038109	0.041286
11			0.11627	0.121692	0.108632	0.116228	0.126823
12		0.21529	0.101541	0.04108	0.045104		

Table 4.16 HiCOPS's write stalls/total stalls (wps) for the 12 experiments (exp_i) for increasing number of parallel nodes (n)

Exp. No.	n = 1	n = 2	n = 4	n = 8	n = 16	n = 32	n = 64
1		0.044665	0.03385	0.02621	0.019666		
2		0.201336	0.184515	0.169855	0.151227		
3	0.113917	0.106945	0.092391	0.079345	0.073518	0.070596	
4			0.13683	0.126815	0.126815	0.126815	0.126815
5		0.136007	0.123383	0.101126	0.095139	0.086466	0.062529
6					0.189123	0.15903	0.136501
9						0.162653	0.139683
10		0.18857	0.161705	0.137424	0.123278	0.107838	0.092693
11			0.123683	0.126815	0.130513	0.125154	0.132481
12		0.213821	0.212703	0.20419	0.186142		

Table 4.17 Percentage HiCOPS's time spent in load imbalance overhead for the 12 experiments (exp_i) for for the 12 experiments (exp_i) increasing number of parallel nodes (n)

Exp. No.	n = 1	n = 2	n = 4	n = 8	n = 16	n = 32	n = 64	n = 72
1		2.247232	5.478383	5.468814	9.327324			
2		0.203377	0.476728	0.751866	1.537653			
3	0.011691	0.022445	0.041688	0.068697	0.312204	0.706813		
4			0.007338	0.013869	0.808574	2.492005	2.61933	
5		0.143813	0.302233	0.039956	0.431971	0.233134	5.375071	
6					0.007387	0.015479	0.024756	
7						0.005527	0.076835	
8							1.594526	1.02223
9						0.004774	0.00895	
10		0.003223	0.008067	0.016665	0.029342	0.051187	0.072798	
11			0.964836	3.017238	3.386828	11.24338	12.52164	
12		0.092163	0.24878	0.5253	0.743586			

Table 4.18 Percentage HiCOPS's time spent in inter-process communication overhead for the 12 experiments (exp_i) for increasing number of parallel nodes (n)

Exp. No.	n = 1	n = 2	n = 4	n = 8	n = 16	n = 32	n = 64	n = 72
1		5.232105	3.926955	7.904505	6.856858			
2		3.79414	2.314373	3.384774	4.295352			
3	0.08221	0.478144	0.930593	1.765939	2.93689	4.200068		
4			0.136043	0.332012	0.88928	4.020877	2.097651	
5		0.016538	0.063838	0.198492	0.411026	0.900368	1.031999	
6					0.431297	0.743349	1.297135	
7						0.495525	1.015206	
8							0.834609	0.643964
9						0.309112	1.056103	
10		0.058962	0.283043	0.505798	2.516395	1.953969	3.919733	
11			0.469823	0.927757	2.55235	1.740178	1.567835	
12		0.4885	1.746925	2.267319	5.660874			

Table 4.19 Percentage HiCOPS's time spent as pipeline halt overhead for the 12 experiments (exp_i) for increasing number of parallel nodes (n)

Exp. No.	n = 1	n = 2	n = 4	n = 8	n = 16	n = 32	n = 64	n = 72
1		2.247232	5.478383	5.468814	9.327324			
2		0.203377	0.476728	0.751866	1.537653			
3	0.011691	0.022445	0.041688	0.068697	0.312204	0.706813		
4			0.007338	0.013869	0.808574	2.492005	2.61933	
5		0.143813	0.302233	0.039956	0.431971	0.233134	5.375071	
6					0.007387	0.015479	0.024756	
7						0.005527	0.076835	
8							1.594526	1.02223
9						0.004774	0.00895	
10		0.003223	0.008067	0.016665	0.029342	0.051187	0.072798	
11			0.964836	3.017238	3.386828	11.24338	12.52164	
12		0.092163	0.24878	0.5253	0.743586			

Table 4.20 Percentage HiCOPS's time spent in I/O overhead for the 12 experiments (exp_i) for increasing number of parallel nodes (n)

Exp. No.	n = 1	n = 2	n = 4	n = 8	n = 16	n = 32	n = 64	n = 72
1		2.535598	3.336975	4.120759	4.190594			
2		1.251304	2.853384	7.482856	13.60012			
3	0.831183	1.611824	4.65796	4.973139	10.05522	10.22056		
4			2.559578	3.584081	7.430818	12.48344	18.53344	
5		0.518943	1.498282	3.304334	5.407848	7.547681	14.42935	
6					0.339751	1.347544	1.12055	
7						0.39289	1.641276	
8							6.90806	7.375666
9						0.427518	1.045725	
10		0.711574	0.586334	1.414206	5.262876	4.497312	6.250695	
11			5.24391	8.263088	7.360256	10.34876	15.43937	
12		0.581338	1.388488	1.955603	3.611701			

References

1. Haseeb M, Saeed F (2021) High performance computing framework for tera-scale database search of mass spectrometry data. Nat Comput Sci 1(8):550–561
2. Nesvizhskii AI, Roos FF, Grossmann J, Vogelzang M, Eddes JS, Gruissem W, Baginsky S, Aebersold R (2006) Dynamic spectrum quality assessment and iterative computational analysis of shotgun proteomic data toward more efficient identification of post-translational modifications, sequence polymorphisms, and novel peptides. Mol Cell Proteomics 5(4):652–670
3. Nesvizhskii AI (2010) A survey of computational methods and error rate estimation procedures for peptide and protein identification in shotgun proteomics. J Proteomics 73(11):2092–2123
4. Chi H, Liu C, Yang H, Zeng WF, Wu L, Zhou WJ, Niu XN, Ding YH, Zhang Y, Wang RM, et al (2018) Open-pfind enables precise, comprehensive and rapid peptide identification in shotgun proteomics. bioRxiv 285395
5. Bern M, Cai Y, Goldberg D (2007) Lookup peaks: a hybrid of de novo sequencing and database search for protein identification by tandem mass spectrometry. Anal Chem 79(4):1393–1400
6. Frank A, Pevzner P (2005) Pepnovo: de novo peptide sequencing via probabilistic network modeling. Anal Chem 77(4):964–973
7. Chi H, Sun R-X, Yang B, Song C-Q, Wang L-H, Liu C, Fu Y, Yuan Z-F, Wang H-P, He S-M et al (2010) pNovo: de novo peptide sequencing and identification using HCD spectra. J Proteome Res 9(5):2713–2724
8. Taylor JA, Johnson RS (2001) Implementation and uses of automated de novo peptide sequencing by tandem mass spectrometry. Anal Chem 73(11):2594–2604
9. Zhang J, Xin L, Shan B, Chen W, Xie M, Yuen D, Zhang W, Zhang Z, Lajoie GA, Ma B (2012) PEAKS DB: de novo sequencing assisted database search for sensitive and accurate peptide identification. Mol Cell Proteomics 11(4):M111-010587
10. Devabhaktuni A, Lin S, Zhang L, Swaminathan K, Gonzalez CG, Olsson N, Pearlman SM, Rawson K, Elias JE (2019) Taggraph reveals vast protein modification landscapes from large tandem mass spectrometry datasets. Nat Biotechnol 37(4):1
11. Kong AT, Leprevost FV, Avtonomov DM, Mellacheruvu D, Nesvizhskii AI (2017) MSFragger: ultrafast and comprehensive peptide identification in mass spectrometry-based proteomics. Nat Methods 14(5):513
12. McIlwain S, Tamura K, Kertesz-Farkas A, Grant CE, Diament B, Frewen B, Howbert JJ, Hoopmann MR, Kall L, Eng JK et al (2014) Crux: rapid open source protein tandem mass spectrometry analysis. J Proteome Res 13(10):4488–4491
13. Yuan ZF, Liu C, Wang HP, Sun RX, Fu Y, Zhang JF, Wang LH, Chi H, Li Y, Xiu LY, et al (2012) pParse: A method for accurate determination of monoisotopic peaks in high-resolution mass spectra. Proteomics 12(2):226–235
14. Deng Y, Ren Z, Pan Q, Qi D, Wen B, Ren Y, Yang H, Wu L, Chen F, Liu S (2019) pClean: an algorithm to preprocess high-resolution tandem mass spectra for database searching. J Proteome Res 18(9):3235–3244
15. Degroeve S, Martens L (2013) MS2PIP: a tool for MS/MS peak intensity prediction. Bioinformatics 29(24):3199–3203
16. Zhou X-X, Zeng W-F, Chi H, Luo C, Liu C, Zhan J, He S-M, Zhang Z (2017) pDeep: predicting MS/MS spectra of peptides with deep learning. Anal Chem 89(23):12690–12697
17. Eng JK, McCormack AL, Yates JR (1994) An approach to correlate tandem mass spectral data of peptides with amino acid sequences in a protein database. J Am Soc Mass Spectrom 5(11):976–989
18. Craig R, Beavis RC (2003) A method for reducing the time required to match protein sequences with tandem mass spectra. Rapid Commun Mass Spectrom 17(20):2310–2316
19. Diament BJ, Noble WS (2011) Faster SEQUEST searching for peptide identification from tandem mass spectra. J Proteome Res 10(9):3871–3879
20. Eng JK, Fischer B, Grossmann J, MacCoss MJ (2008) A fast SEQUEST cross correlation algorithm. J Proteome Res 7(10):4598–4602

21. Park CY, Klammer AA, Kall L, MacCoss MJ, Noble WS (2008) Rapid and accurate peptide identification from tandem mass spectra. J Proteome Res 7(7):3022–3027
22. Williams S, Waterman A, Patterson D (2009) Roofline: an insightful visual performance model for multicore architectures. Commun ACM 52(4):65–76
23. Chi H, He K, Yang B, Chen Z, Sun R-X, Fan S-B, Zhang K, Liu C, Yuan Z-F, Wang Q-H et al (2015) pFind-Alioth: a novel unrestricted database search algorithm to improve the interpretation of high-resolution MS/MS data. J Proteomics 125:89–97
24. Marx V (2013) Biology: the big challenges of big data
25. Duncan DT, Craig R, Link AJ (2005) Parallel tandem: a program for parallel processing of tandem mass spectra using PVM or MPI and x! tandem. J Proteome Res 4(5):1842–1847
26. Bjornson RD, Carriero NJ, Colangelo C, Shifman M, Cheung K-H, Miller PL, Williams K (2007) X!! tandem, an improved method for running x! tandem in parallel on collections of commodity computers. J Proteome Res 7(1):293–299
27. Pratt B, Howbert JJ, Tasman NI, Nilsson EJ (2011) MR-tandem: parallel x! tandem using hadoop MapReduce on amazon web services. Bioinformatics 28(1):136–137
28. Li C, Li K, Chen T, Zhu Y, He Q (2019) SW-Tandem: a highly efficient tool for large-scale peptide sequencing with parallel spectrum dot product on Sunway TaihuLight. Bioinformatics (Oxford, England) 35(19):3861–3863
29. Li C, Li K, Li K, Lin F (2019) MCtandem: an efficient tool for large-scale peptide identification on many integrated core (MIC) architecture. BMC Bioinformatics 20(1):397
30. Prakash A, Ahmad S, Majumder S, Jenkins C, Orsburn B (2019) Bolt: a new age peptide search engine for comprehensive MS/MS sequencing through vast protein databases in minutes. J Am Soc Mass Spectrom 30(11):2408–2418
31. Chen L, Zhang B, Schnaubelt M, Shah P, Aiyetan P, Chan D, Zhang H, Zhang Z (2018) MS-PyCloud: an open-source, cloud computing-based pipeline for LC-MS/MS data analysis. bioRxiv 320887
32. Kim S, Pevzner PA (2014) MS-GF+ makes progress towards a universal database search tool for proteomics. Nat Commun 5(1):5277
33. Kulkarni G, Kalyanaraman A, Cannon WR, Baxter D (2009) A scalable parallel approach for peptide identification from large-scale mass spectrometry data. In: 2009 international conference on parallel processing workshops. IEEE, pp 423–430
34. Valiant LG (1990) A bridging model for parallel computation. Commun ACM 33(8):103–111
35. Haseeb M, Afzali F, Saeed F (2019) LBE: a computational load balancing algorithm for speeding up parallel peptide search in mass-spectrometry based proteomics. In: IEEE international parallel and distributed processing symposium workshops (IPDPSW). IEEE 2019, pp 191–198
36. Madsen JR, Awan MG, Brunie H, Deslippe J, Gayatri R, Oliker L, Wang Y, Yang C, Williams S (2020) Timemory: modular performance analysis for HPC. In: International conference on high performance computing. Springer, pp 434–452
37. Stevens R, Ramprakash J, Messina P, Papka M, Riley K (2019) Aurora: argonne's next-generation exascale supercomputer. Technical report, ANL (Argonne National Laboratory (ANL), Argonne, IL (United States))
38. Liu K, Li S, Wang L, Ye Y, Tang H (2020) Full-spectrum prediction of peptides tandem mass spectra using deep neural network. Anal Chem 92(6):4275–4283
39. Lin Y-M, Chen C-T, Chang J-M (2019) MS2CNN: predicting MS/MS spectrum based on protein sequence using deep convolutional neural networks. BMC Genomics 20(9):1–10

Chapter 5
Fast Spectral Pre-processing for Big MS Data

Fahad Saeed and Muhammad Haseeb

In this chapter, we discuss and introduce a data pre-processing algorithm for dimensionality reduction of big MS data. We will start by discussing a few spectral pre-processing methods followed by the introduction MS-REDUCE which is a highly efficient method for processing MS data

5.1 A Review of Spectral Pre-processing Methods

One of the most essential and fundamental processes that is used for MS-based omics data needs processing of the data so that it can be fed to the available tools. Pro-processing of spectral data is essential for MS-based omics data in recent years. This pre-processing of MS data has an objective, i.e. to allow readability of MS data using tools that are available to process it, e.g. Tide, MSFragger.

Pre-processing for MS data can include a plethora of method and some of them are listed below:

- Spectral Clustering [2];
- Noise Reduction in Spectra [3];
- Quality assessment of spectra [4];
- Precursor charge determination [5].
- b-y Ion Separation [6].

The objective of these methods is to reduce the noise variance in the spectra, which may lead to better peptide identification (with better confidence in the deduced peptides), and perhaps reduce the amount it takes for the tools to complete the processing. Since spectral noise reduction is the most readily used technique for

---

Some parts of this chapter may have appeared in [1].

F. Saeed and M. Haseeb, *High-Performance Algorithms for Mass Spectrometry-Based Omics*, Computational Biology 34,
https://doi.org/10.1007/978-3-031-01960-9_5

processing of MS data, we introduce a few algorithms that have been proposed to date.

5.1.1 Spectral Denoising Algorithms

There are multiple spectral denoising algorithms that have been proposed till date with various objectives. Most of the algorithms try to deduce which peak is noise and correspondingly remove it or increase the intensity of the peak measured as a function of the spectra under consideration.

5.1.1.1 MS-Cleaner Algorithm

MSCleaner [7] removes the unwanted peaks from the spectra to reduce noise and data that needs to be processed by the search engines. Four different algorithms were proposed which included numerical analysis and signal processing techniques which allowed the method to look for charged ions, isotopic clusters, background noise, and, more importantly, detection of spectra that are not interpretable. Although the methods are reported to work well for a variety of spectra, these methods are deemed too compute-intensive for processing of big MS data. In this study, the authors have reported the optimized version of their implementation to process spectra in 0.25 s for a data set that is 53944. Of course, the current rate at which MS data can be produced is much more than a few thousands making processing using these traditional methods less than ideal. However, with increased computational load the authors were able to show a reduction of 15%–39% in raw data. MS Cleaner software, a version 2.0 in 2010 [8] was also introduced by the same subset of authors. This improved version detects the peptide ladder sequence using a fixed number of most intense peaks from each spectrum and results in a faster reduction in data by up to 80% as reported by the authors. The improved speed of the methods also results in about 0.02s–0.08 s per spectrum depending on the kind of spectra being processed.

Ding's Denoising Algorithm:

Another denoising algorithm [3] consists of two steps: (1) peak intensity is adjusted using five different features. In the later stage, a morphological reconstruction filter is employed to remove the noisy peaks based on their adjusted intensity in the previous stage. The method is reported to reduce the data by up to 60% but our experiments have demonstrated that the processes are extremely compute-intensive with results of 1 million spectra taking more than 3 days to complete processing. Two other similar methods are also published [9, 10] but suffer from the same problem of compute-intensive workflows—For example, [10] takes approximately 1.7 s per spectrum and would take an exceedingly long time to process millions of spectra.

5.1.2 Spectral Quality Assessment Algorithms

As a pre-processing step, there are multiple studies that have shown that assessment of quality of the spectra can also contribute to better and more accurate peptide deduction. For example, the authors in this study [11] estimate the quality of the spectra by treating the problem as a constraint optimization problem and show that up to 74% of the low-quality spectra could be removed using this technique. They also show that almost 10% of the spectra is also eliminated as a result of this method. In order to remove this limitation, authors in [12] have introduced a novel feature that is based on cumulative intensity normalization and shows that 60% of the low-quality spectra can be removed while losing 2% of the high-quality spectra. Other methods include [4, 13–15] which use various technique to eliminate the low-quality spectra. However, most of these proposed techniques are very compute-intensive which makes them infeasible for big data MS-based omics.

5.1.3 Separation of b-y Ions

As discussed in this chapter, the elimination of low-quality spectra can help in pre-processing phase of MS data analysis. However, as expected if we are eliminating low-quality spectra; all the useful information that might be in the spectra is also eliminated. Therefore, there are multiple methods that try to eliminate the peaks that might be of low quality instead of eliminating the whole spectra. Since the existing algorithm implicitly assumes that all the b- and y-ions are present, using such pre-processing technique necessitates that post-processing algorithms be redesigned so that the method can incorporate using only b- and y-ions without the explicit assumptions that all peaks are available.

We are aware of only two algorithms that operate by removing incorrect or noisy b- or y-ions from MS^2 spectrum. One such method [6] uses a graph-theoretic model, constructs a de novo spectra graph [16] and uses edge-weights for other features of the spectra to partition the graph into multiple smaller subgraphs. These subgraphs are then labeled as sets of b- and y-ions based on a final graph after the partitioning which can be used for further processing. This method is shown to give impressive results on *only* simulated data sets, and very high-resolution high-quality spectra. However, the experiments for experimental data are highly limited and have been performed over a very small number of high-quality experimental spectra. Another tool known as MS-REDUCE is also shown to give impressive results on simulated, and real data sets on a wide variety of parameters and is shown to be low-computational complexity which makes it ideal for processing big MS data. MS-REDUCE methods and their parallelization is discussed in the next few chapters.

5.2 MS-REDUCE: An Ultra-Fast Data Reduction Algorithm for Big MS Data

We introduce highly efficient data processing algorithm for dimensionality reduction of big MS data called MS-REDUCE [1, 17] which allows a reduction in the number of peaks, and in turn the overall data that needs to be processed by subsequent workflows. The proposed MS-REDUCE method is a low-computational solution that operates by selecting and sampling peaks based on pre-defined quantized levels. Note that all the peaks that are selected to be useful peaks are determined *before* any peptide deduction calculations.[1] To make the process of low-computational intensity, all the calculations are performed on only a handful (of sampled) of peaks irrespective of the size of the spectra under consideration. Due to this reason, MS-REDUCE can operate in linear time with respect to the size of the spectra. Here, we formally introduce the problem. Notations will be introduced and defined wherever they occur first throughout the manuscript.

Definition 5.1 Let there be N number of spectra $S = \{s_1, s_2, \ldots, s_N\}$. If length of spectrum s_i is l_i, then each spectrum can be represented as a series of peaks, i.e. $s_i = \{p_1, p_2, p_3, \ldots, p_l\}$, where p is a peak in a spectrum.

Definition 5.2 If s_i' denotes a spectrum after being processed by MS-REDUCE and the size of the processed spectrum be l_i', then R is the reduction factor such that $R = (l_i'/l_i) * 100$ for each spectrum.

Each spectrum s in S has to reduced to s' so that both s and s' belong to the same peptide with some degree of confidence. There are of course cases where s and s' belong to different peptides. In that case, if the peptide match for s' has a confidence value better than the threshold value to qualify for a high-confidence hit, then that counts as a correct hit. One example of such an instance is assumed that there is a spectrum s that was incorrectly mapped to peptide A. However, after MS-REDUCE processing and the resulting suppression in noise levels, the spectrum s' may correspond correctly to another peptide B or vice versa. Correctness is determined by the quality assessment that we will discuss in this chapter.

The proposed solution MS-REDUCE exploits the data that is obtained from the MS instruments, i.e. 90% of peaks in given spectra are noise or that they may not be useful when deducing peptides [8]. MS-REDUCE also exploits the variation in the MS data and runs using a three-stage pipeline as shown in Fig. 5.1: (1) Spectral Classification; (2) Peak Quantization; (3) Weighted Random Sampling.

The spectral classification module is the first stage and its objective is to determine the spectrum noise levels of a given spectrum. Our earlier paper [1, 17] has introduced a novel metric, called Spectral Intensity Spread (SIS), that roughly estimates the diversity (or deviation) of the peaks in a given data set. This part of the method operates on the assumption that the larger the value of SIS, the more noisy a given spectrum is, and hence more noise needs to be eliminated [18].

---

[1] Knowing which peaks are useful after the peptide deduction is a trivial operation.

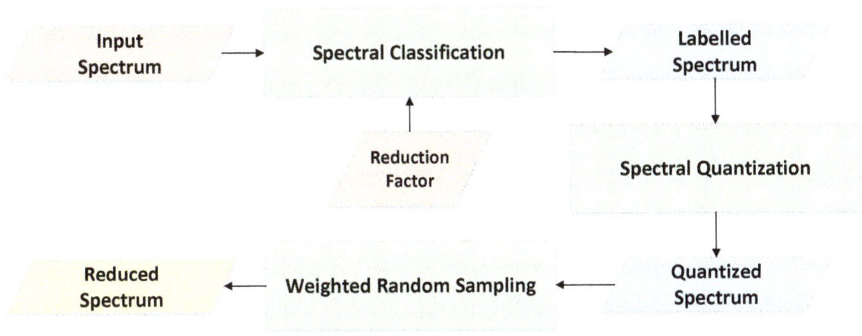

Fig. 5.1 Figure showing the pipeline for MS-REDUCE algorithm

After the first phase of putting the spectra in the correct SIS class, the next module is called spectral quantization module where the spectra are quantized using different levels of intensity. The number of quantization levels depends on the class of spectra that was assigned in the previous stage. If the spectra are more noisy, more quantization levels are assigned to the spectra. Since the module distributes the peaks into different groups based on the intensity levels of the spectra, therefore, it is much simpler and faster to access the peaks based on their intensity levels.

The last module takes the quantized spectra where the signal peaks are randomly selected on the quanta. The number of peaks that are retained is dependent on the user-defined reduction factor R. Weighted sampling rates are calculated for each quantum so that summation of the peaks collected at each level is equal to the percentage of the peaks that are required. Sampling rate is the percentage of the peak that is retained in each quantization level. Below we describe each of the modules in detail.

The number of peaks to be retained is calculated based on the user-defined reduction factor R.

5.2.1 *Spectral Classification*

Most of the previous pre-processing algorithms that deal with noise reduction of MS data, considered all spectra in the same class, i.e. MS spectra that have a better quality were treated in the same way as spectra that had a poor signal-to-noise ratio. This assumption results in compute-intensive tasks for spectra which may not need much processing (due to being better quality spectra). In this module, MS-REDUCE classifies MS data based on the quality and S/N ratio of the spectra under consideration. The classification takes place by classifying the spectra using an approximation of the noise content in them.

5.2.1.1 Intensity Spread

The classification of the spectra takes place by comparing the spectrum's intensity spread with the average intensity spread of the whole data set under consideration. With this formulation, no matter what the quality of the MS data is in a given data set, the methodology always classifies the spectra between spectra that are comparatively better in quality than other spectra under consideration.

More formally:

Definition 5.3 Let N be the total number of spectra in set S then $S = \{s_1, s_2, s_3, \ldots, s_n\}$, where s_i represents one spectrum. Then the intensity spread for spectrum s_i can be calculated as

$$V_i = \text{Max10Avg}(s_i) - \text{Min10Avg}(s_i) \tag{5.1}$$

where V_i is the intensity spread of the spectrum i, and $\text{Max10Avg}(s_i)$ and $\text{Min10Avg}(s_i)$ present the average of ten most and least intense peaks of the spectrum, respectively. Similarly, average intensity spread for a data set can be calculated as

$$V_{\text{avg}} = \frac{\sum_{i=1}^{N}(\text{Max10Avg}(s_i) - \text{Min10Avg}(s_i)}{N} \tag{5.2}$$

where: V_{avg} = Average Intensity Spread N = number of spectra in set S

As expected, this classification is a very low-complexity procedure and as dictated by Eq. (5.2), the calculation requires processing of only 20 peaks per spectra regardless of its size.

5.2.1.2 Classification

All spectra are classified into four different classes and the grouping takes place by calculating how much below or above the value of V is with respect to the V_{avg}. Any classes that are named are in increasing numerical order and higher classes contain spectra with larger value V and vice versa. The threshold values V are also assigned based on the data sets V_{avg} that would allow mapping to a particular class.

More formally, the threshold values for each class can be defined as follows:

Definition 5.4 Let x denote a class then for x = {1, 2, 3}

$$S_x = \left\{ s_i | (x-1) * \frac{1}{4} * V_{avg} \leq V_i \leq x * \frac{1}{4} * V_{\text{avg}} \right\} \tag{5.3}$$

and for x = {4}:

$$S_x = \left\{ s_i | \frac{3}{4} * V_{\text{avg}} \leq V_i \right\} \tag{5.4}$$

where: S_x = Class x containing spectra assigned to it.

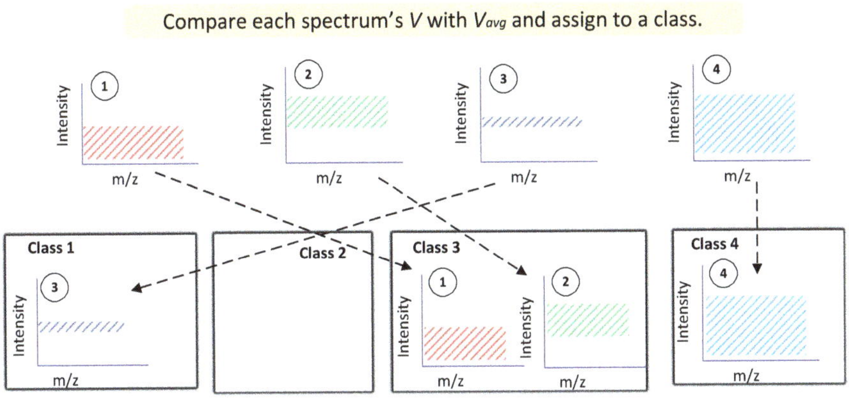

Fig. 5.2 Figure depicts a visual representation of classification stage. The shaded regions present the Spectral Spread (V); larger the shaded area, the larger the value of V and noisier the corresponding spectrum is considered

Figure 5.2 shows an example of two spectra (1 and 2) which have a varied range of intensities. However, their spectra spread is almost the same and, therefore, they have been assigned to the same class.

5.2.2 *Spectral Quantization*

The quantization of the spectra takes place along the intensity axis. The intensity of the peaks are compared with the upper and lower quanta of the spectra that have been defined, and if these comparisons are within a certain limit, they are assigned to that quantum. This process allows us to have separate bins which contain peaks within a specific range of intensities. The useful peaks are then picked from their quanta and added to the final reduced spectra as explained in the next step. This also allows the method to make decisions without heavy computations.

5.2.2.1 Quantization Levels

The number of quantization levels is chosen so the spectra that have wide intensity spread are quantized in a large number of levels, and narrower spread is divided into smaller number of levels. An example of quantization of class I spectrum is illustrated in Fig. 5.3. Our preliminary experiments have suggested that a spectrum with a smaller intensity spread does not improve the quality of the spectra but contributes to increasing the processing time. Smallest number of quantization levels are used to reduce time- and space requirements. Our proposed method assigns 1, 2, 3, and 4 classes are 5, 7, 9, and 11 levels of quantization, and these numbers are obtained by extensive empirical observations and experiments.

The quantization process can be formally defined as

Definition 5.5 Let n_x be the maximum number of quantization levels for class x, then we can have $n_1 = 5$, $n_2 = 7$, $n_3 = 9$, and $n_4 = 11$. q_{ij} represents the quantum j of spectrum i. Then following equations are calculated for each spectrum s_i, for each quantum j from 1 till n_x for $j < n_x$

$$q_{ij} = \{p| \frac{(j-1)}{n_x} * \mathrm{M10A}(s_i) \leq \|p\| \leq \frac{j}{n_x} * \mathrm{M10A}(s_i)\} \tag{5.5}$$

for $j = n_x$

$$q_{ij} = \{p| \frac{(j-1)}{n_x} * \mathrm{M10A}(s_i) \leq \|p\|\} \tag{5.6}$$

where: j = quantization level under consideration q_{ij} = jth quantization level of ith spectrum n_x = number of quantization levels for class x $\|p\|$ = intensity of peak p $\mathrm{M10A}(s_i)$ = Average Intensity of 10 most intense peaks of s_i

Equations (5.5) and (5.6) are computed for each value of n_x ranging from 1 till n_x.

The quantum number that is assigned to each peak represents certain characteristics, e.g. any of the peaks in quantum 1 are the lowest peaks and contain the least intense peaks that are available in that data set. Similarly, quantum 11 is the highest quanta level for class 4 spectra and would contain peaks with the highest intensity. By design, we have made sure that the quanta are equally spaced because 90% of data is redundant, so the probability that any outlier (if any) will affect the quality of the hits is extremely small and any irregular quantization is likely not needed.

5.2.3 *Weighted Random Sampling*

Probability of finding a peak that is useful is greater in higher quanta [19]. The underlying assumption of this module is that each peak within one quantization level has equal probability of being useful for peptide deduction. Therefore, in order to determine which peaks must be sampled from each quantum, sampling weights are computed which are explained below (Fig. 5.4).

5.2.3.1 Weights Calculation

The number of peaks that are retained have to be calculated given by a reduction factor by the user. The recursive method estimating the sampling weights for each of the defined quantum has to satisfy the equation below:

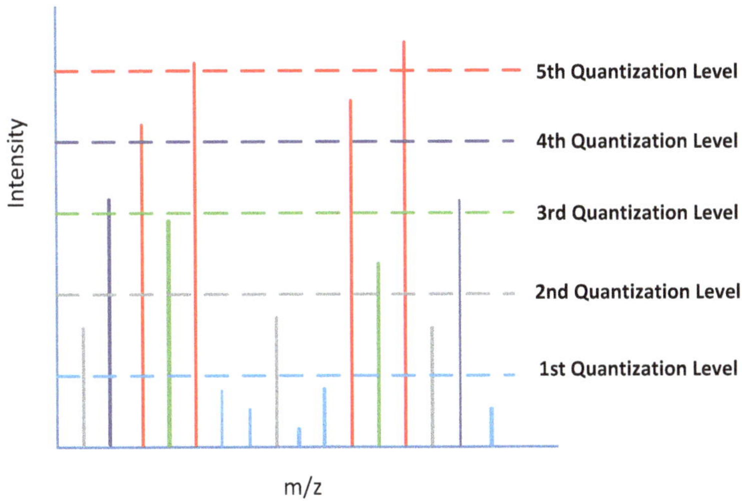

Fig. 5.3 Figure represents quantization of a class I spectrum with five different quantization levels. The red-colored peaks are the most intense and belong to the fifth quantum, while the light blue-colored are the least intense and have been binned into the lowest quantum

$$\sum_{i=1}^{n_x}\left(\frac{x_i}{100} * q_i\right) = p^{'} \tag{5.7}$$

where: x_i = sampling rate for quantization level i $q_i = i_{\text{th}}$ quantization level $\|q_i\|$ = number of peaks at i_{th} quantization level $p^{'}$ = number of peaks required to satisfy the reduction factor

Peaks are then selected from the highest quantization levels and then continued downwards till the required number of peaks are reached. If there are more peaks than that are required at a given quantization level. random sampling of the peaks is done to select the required number of peaks.

Formally, this can be presented by Eqs. (5.8) through (5.10): case 1: $\|q_{n_x}\| = p^{'}$

$$x_i = \begin{cases} 100, & \text{if } i = n_x. \\ 0, & \text{otherwise} \end{cases} \tag{5.8}$$

case 2: $\|q_{n_x}\| > p^{'}$

$$x_i = \begin{cases} \frac{\|q_i\| - (\|q_i\| - p^{'}}{\|q_i\|}), & \text{if } i = n_x. \\ 0, & \text{otherwise} \end{cases} \tag{5.9}$$

case 3: Default

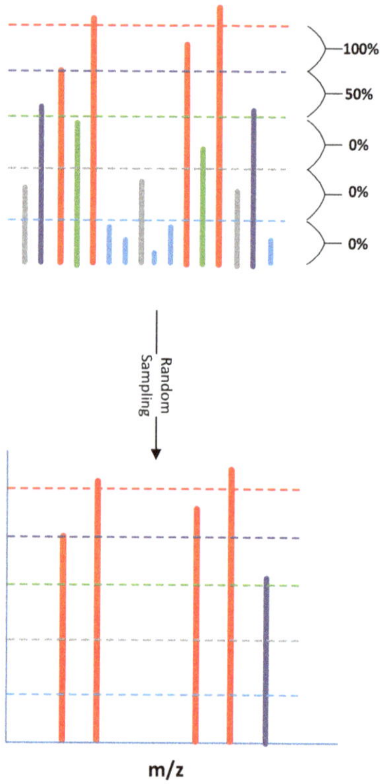

Fig. 5.4 Figure presents a visual representation of the random sampling module. In this figure, the topmost quantum is assigned a weight of 100%, while the fourth quantum is assigned a weight of 50%. Peaks from all other quanta are discarded owing to their zero sampling rate

$$x_i = \begin{cases} 100, & \text{if } p' - \|q_j\| > \|q_{j+1}\|. \\ \frac{p' - \sum_{j=i+1}^{n_x} \|q_j\|}{\|q_i\|}, & \text{otherwise} \end{cases} \tag{5.10}$$

Figure 5.4 shows an example of weighted random sampling executed on class I spectra. The figure on the right shows the reduced spectra and the two peaks that are chosen from the fourth quanta only one of them appears in the final spectrum which has been randomly chosen.

5.3 Performance Evaluation of MS-REDUCE

Performance evaluation of MS-REDUCE is done in the following two phases: (1) execution time and speedups are observed and compared against the existing algorithms; (2) Quality assessment of data produced as a result of MS-REDUCE is done, i.e. the quality of the matches after the reduction in the data is done. All of the experiments were performed using experimentally obtained data sets. Reduction in the data and the accuracy of the peptide deduction are compared with the existing tools [17].

5.3.1 Time Complexity

Time complexity of MS-REDUCE can be computed by observing the big-O complexity of each of the modules and then summing them. We show below that this makes MS-REDUCE having a linear time complexity with respect to the number of spectra.

Time complexity calculations are as follows:

- First module is Intensity Spread which involves two processes. First is the sorting of spectra and second step is calculating the average intensity of max and min peaks.The first step is a $O(l_i log(l_i))$ process where l_i is the length of spectrum i. The second step is a constant time process and has a time complexity of $O(1)$. This is done for each spectrum which makes the total time complexity equal to N, i.e. $O(N)$ assuming that the total number of spectra to be N, where $N >> l_i$.
- Second step classifies spectra into classes depending on the noise level associated with their level. This requires a comparison of Average Intensity Spread for each spectra. The comparison step is a constant time process. Repeating this for N spectra results in linear time complexity of $O(N)$, where N is the number of spectra.
- Quantization module requires comparison of each peak that has pre-calculated values repeated 4 times and is a constant time process. This results in time complexity of $O(4 * l)$. Module works on each spectra giving a time complexity of $O(4 * l * N)$ as N is much larger than other values so it can be presented as $O(N)$.
- The last module performs a random sampling of peaks based upon user-defined sampling rate R. Number of peaks to be sampled is given by $s * l$. For N spectra time complexity is equal to $O(s * l * N)$. This also results in $O(N)$ time complexity.

Result can be summed up to $O(4 * N * L)$ which is approximately equivalent to $O(N)$ since N is much larger than L.

5.3.2 Experimental Verification of the Complexity Analysis

After the theoretical computational complexity, experimental assessment and verification are needed. In order to accomplish this, we used data sets of various qualities, experiments, and numbers. For all the experiments, we replicated the UPS2 data sets as many times as was needed to get the desired sizes. Ten different data sets, each having 100,000 spectra were produced and all the experiments were performed using Linux-based server with 24 cores, each operating at 1200 MHz. For our single-threaded implementation of MS-REDUCE, the result shown in Fig. 5.5 depicts linear

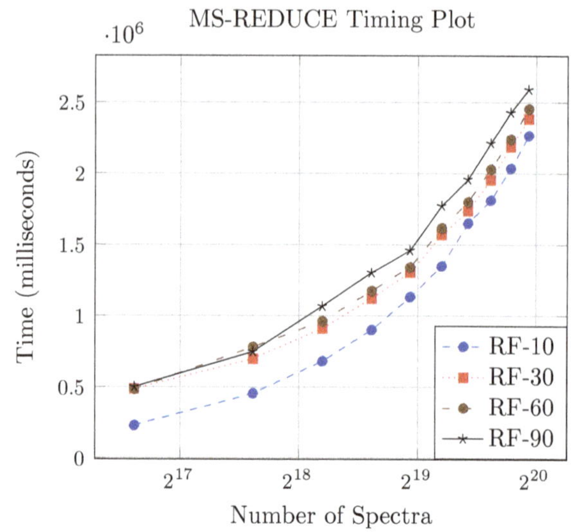

Fig. 5.5 Figure showing a graph between processing time of MS-REDUCE and the number of spectra processed at reduction factors of 10, 30, 60, and 90. The horizontal axis represents the number of spectra, while the vertical axis represents time in milliseconds

time complexity with respect to the number of spectra which concurs with our theoretical computational complexity analysis. All experiments were performed with user-defined reduction factors 10, 30, 60, and 90, and in order to compensate for system irregularities each experiment was performed 10 times, and average execution time was reported. Reduction factor as defined previously is the amount of data that is retained by the algorithm, and our execution results are shown in Fig. 5.5 clearly show that the change in the reduction factor does not affect the run time significantly, and retains almost linear-trend for various levels of reduction-factors.

5.3.3 Speed Comparison

The proposed MS-REDUCE was also compared with other leading start-of-the-art denoising methods [3]. To accomplish a fair comparison, two metrics were defined: (1) Conventional speedup (i.e. how fast or slow is one method from the other); (2) Spectra per second (SPS) that is analyzed using the said method. Below we define both metrics in a more formal way:

$$S = T_{other}/T_{reduce} \tag{5.11}$$

Here, S is the speed up obtained, T_{other} is the processing time of competing method, and T_{reduce} is the time taken by MS-REDUCE.

$$SPS = Spectra/Time \tag{5.12}$$

Table 5.1 Speedup achieved by MS-REDUCE over the denoising algorithm

Spectra	$T_{denoise}$(msec)	$T_{ms-reduce}$(msec)	Speed up
9.61×10^4	2.35×10^7	2.25×10^5	103
1.92×10^5	4.41×10^7	4.50×10^5	96
2.89×10^5	6.49×10^7	6.78×10^5	94
3.85×10^5	8.60×10^7	9.03×10^5	94
4.81×10^5	1.09×10^8	1.12×10^6	95
5.78×10^5	1.31×10^8	1.75×10^6	83
6.74×10^5	1.55×10^8	2.0×10^6	86
7.71×10^5	1.76×10^8	2.05×10^6	94
8.67×10^5	1.97×10^8	2.29×10^6	94
9.63×10^5	2.20×10^8	2.47×10^6	98
1.06×10^6	2.43×10^8	2.81×10^6	99

Equation (5.11) mathematically shows the calculation for conventional speedup and Eq. (5.12) gives the equation for presents *spectra per second* metric. As can be seen that bigger number for each metric is better for MS-REDUCE, [3] is represented as Denoising Algorithm.

5.3.4 Comparing MS-REDUCE with Other Denoising Methods

Execution of both algorithms was done using similar computational environments to ensure a fair comparison. Table 5.1 shows the results from timing experiments for both MS-REDUCE, and Denoising method [3]. Columns 2, and 3 are shown the execution time in milliseconds. As can be seen De-Noising algorithm takes more than 3 days while MS-REDUCE takes 47 min to process 1 million spectra. This gives MS-REDUCE almost 100 times more speedup as compared to other leading methods. This also ensures that such denoising algorithms can be used for pre-processing of MS data, something that has not been done due to excessively long processing times for these denoising methods.

5.3.5 Quality Assessment

We now discuss the quality of the spectra that is obtained when processed using MS-REDUCE and other leading methods. Figure 5.6 show the workflow that is used to fairly assess the quality of the spectra that is obtained from denoising algorithms. First the raw spectra are fed into the denoising algorithm (MS-REDUCE, and others), and

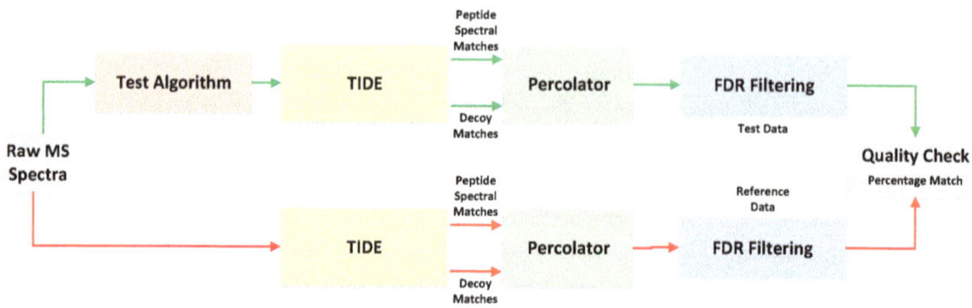

Fig. 5.6 Figure shows the flow of quality assessment experiments. The Test Algorithm shown in the top right corner is replaced by the algorithm under observation i.e. MS-REDUCE or Denoising Algorithm

the spectra that is emitted from these algorithms are sent to Tide [20] search engine of Crux toolkit [21]. Tide provides peptide spectral matches, and decoy matches using a combination of real+decoy databases. Thereafter, percolator [22] does some pre-processing to compute the statistical confidence value for the PSMs and calculated the false discovery rate (FDR), and assigns it to each PSM. We calculated the number of PSMs for same FDR threshold obtained by using the data sets which had been treated by the test algorithm. Using this information it is easy to calculate the percentage of high-quality PSMs obtained by processed spectra (by denoising algorithm), and the PSM obtained when only the raw spectra is used. All experiments that are reported were repeated for FDR values of 1%, 3%, 5%, 7%, and 9%. However, for the depiction of results and making it easier for visualization, we will be using FDR of 5% as a nominal value used for reporting results.

5.3.6 Comparison with Random Sampling of Peaks

We performed the computational quality assessment experiments with all thirteen UPS2 data sets. For ease of visualization only FDR value of 5% is used for results shown in Figs. 5.7, 5.8, 5.9, 5.10 and 5.11 which depict the experiments performed using MS-REDUCE and random peak sampling [17]. For these 3 HCD and 3 CID experiments for UPS2 data sets are reported.

The results graphs have been plotted by varying the reduction factor (from 10% to 90%) of MS-REDUCE, as well as the sampling rate of random peaks selection methods. The 100% would represent that no peaks have been eliminated from the raw spectra and is thus untreated. In general, it is clear that MS-REDUCE gives superior performance by exhibiting percentage matches of around 90% with data reduction rate of *only* 20% i.e. the MS-REDUCE will give highly accurate results even when the number of peaks that are available to process the data as compared to the raw data is only of 20% value. This is a significant result since these results show that 80% of the peaks that are present in the raw data is useless for any peptide deduction (and hence

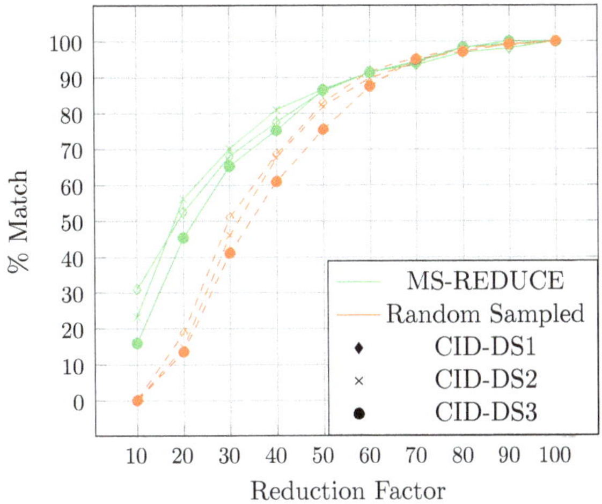

Fig. 5.7 Quality assessment plots for CID-DS1, CID-DS2, and CID-DS3 data sets

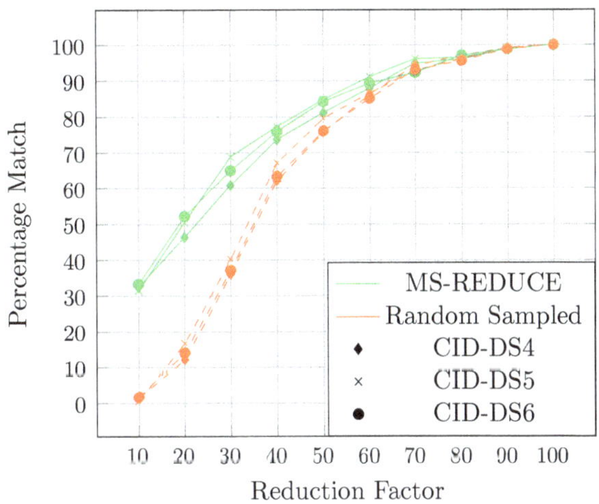

Fig. 5.8 Quality assessment plots for CID-DS4, CID-DS5, and CID-DS6 data sets

can be eliminated). These results are shown to be consistent across experimental conditions, fragmentation types, etc., with overall HCD fragmentation more accurate as compared to CID due to better S/N ratio for HCD data sets [23].

5.3.7 Comparison with Conventional Algorithms

In the last step of the quality assessment process, we compared the quality results of MS-REDUCE as compared to other noise-reducing algorithms. The workflow that is followed for a fair comparison is shown in Fig. 5.6. As can be seen in Fig. 5.12 the quality assessment plots were compared for De-Noising Algorithm, MSCleaner

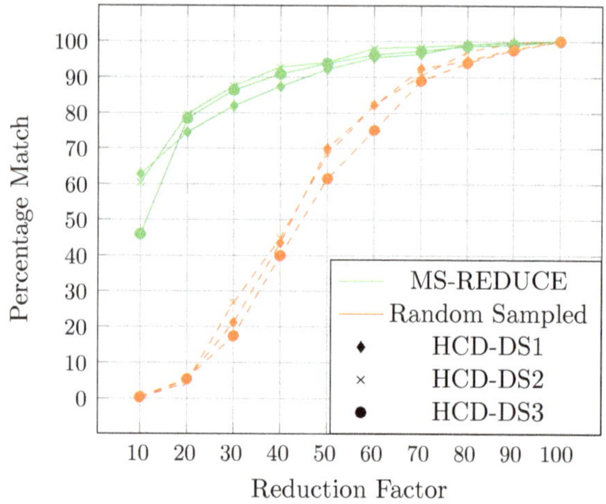

Fig. 5.9 Quality assessment plots for HCD-DS1, HCD-DS2, and HCD-DS3 data sets

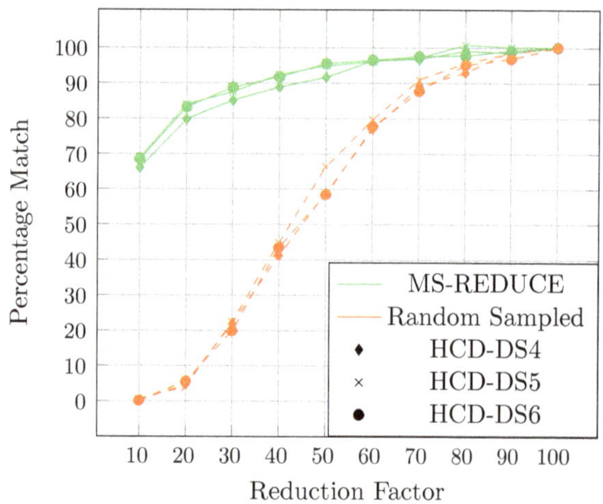

Fig. 5.10 Quality assessment plots for HCD-DS4, HCD-DS5, and HCD-DS6 data sets

2.0, and MS-REDUCE (30, 60, and 90 reduction factors). As these figures show, MS-REDUCE gives a superior performance as compared to MSCleaner 2.0 for all data sets except UPS2 for all reduction factor values. MS-REDUCE also gives comparable performance to Denoising Algorithm when operating at a reduction factor of 60. However, the superior scalability of MS-REDUCE and low-cost computational complexity have a distinct advantage that is not shared with other state-of-the-art denoising techniques.

Even with almost linear-complexity of MS-REDUCE, we would want to take advantage of multicore, manycore, and CPU-GPU architectures that are now ubiquitous. This will be the subject of our next chapters.

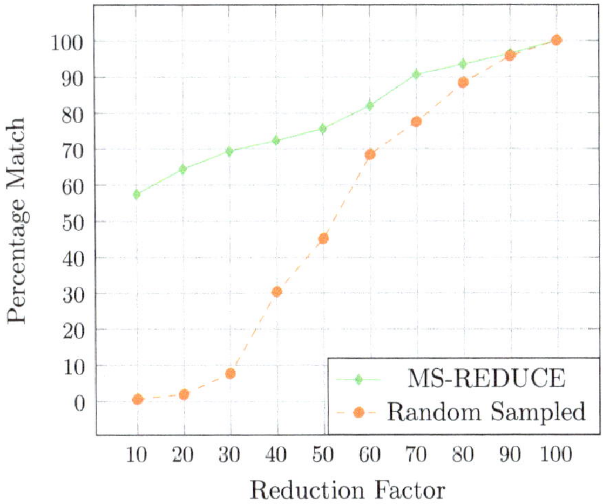

Fig. 5.11 Quality assessment plot for UPS2 data set

Fig. 5.12 Quality assessment plots for MS-REDUCE and other pre-processing algorithms. X-axis shows experimental data sets

References

1. Awan MG, Saeed F (2016) Ms-reduce: an ultrafast technique for reduction of big mass spectrometry data for high-throughput processing. Bioinformatics 32(10):1518–1526
2. Saeed F, Hoffert JD, Knepper MA (2013) Cams-rs: clustering algorithm for large-scale mass spectrometry data using restricted search space and intelligent random sampling. IEEE/ACM Trans Comput Biol Bioinform 11(1):128–141
3. Ding J, Shi J, Poirier GG, Wu F-X (2009) A novel approach to denoising ion trap tandem mass spectra. Proteome Sci 7(1):9
4. Bern M, Goldberg D, McDonald WH, JRY IIII (2004) Automatic quality assessment of peptide tandem mass spectra. Bioinformatics 20
5. Wu F-X, Ding J, Poirier GG (2008) An approach to assessing peptide mass spectral quality without prior information. Int J Funct Inform Pers Med 1(2):140–155
6. Yan B, Pan C, Olman VN, Hettich RL, Xu Y (2004) A graph-theoretic approach for the separation of b and y ions in tandem mass spectra. Bioinformatics 21(5):563–574
7. Mujezinovic N, Raidl G, Hutchins JRA, Peters J-M, Mechtler K, Eisenhaber F (2006) Cleaning of raw peptide ms/ms spectra: improved protein identification following deconvolution of multiply charged peaks, isotope clusters, and removal of background noise. Proteome Sci 6:5117–5131
8. Mujezinovic N, Schneider G, Wildpaner M, Mechtler K, Eisenhaber F (2010) Reducing the haystack to find the needle: improved protein identification after fast elimination of non-interpretable peptide ms/ms spectra and noise reduction. BMC Genomics 11
9. Zhang J, He S, Ling2 CX, Cao X, Zeng R, Gao W (2008) Peakselect: preprocessing tandem mass spectra for better peptide identification. Rapid Commun Mass Spectrom 22
10. Gentzel M, Kocher T, Ponnusamy S, Wilm M (2003) Preprocessing of tandem mass spectrometric data to support automatic protein identification. Proteomics 3
11. Lin W, Wang J, Zhang WJ, Wu FX (2012) An unsupervised machine learning method for assessing quality of tandem mass spectra. Proteome Sci 10
12. Na S, Paek E (2007) Quality assessment of tandem mass spectra based on cumulative intensity normalization. J Proteome Res 5(12)
13. Tabb DL, MacCoss MJ, Wu CC, Anderson SD, III JRY (2003) Similarity among tandem mass spectra from proteomic experiments: detection, significance, and utility. Anal Chem 75(10)
14. Purvine S, Kolker N, Kolker E (2004) Spectral quality assessment for high-throughput tandem mass spectrometry proteomics. OMICS J Integr Biol 8(3)
15. Ding J, Shi J, Wu FX (2011) Svm-rfe based feature selection for tandem mass spectrum quality assessment. Int J Data Min Bioinform 5(1)
16. Dancik V, Addona TA, Clauser KR, Vath JE, Pevzner PA (1999) De novo peptide sequencing via tandem mass spectrometry. J Comput Biol 6(3–4):327–342
17. Awan MG, Saeed F (2015) On the sampling of big mass spectrometry data. In: Proceedings of the 7th international conference on bioinformatics and computational biology, BICOB, pp 143–148
18. Wells G, Prest H, RussIV CW (2011) Why use signal-to-noise as a measure of ms performance when it is often meaningless? Technical report, Agilent Technologies
19. Havilio M, Haddad Y, Smilansky Z (2003) Intensity-based statistical scorer for tandem mass spectrometry. Anal Chem 75(3):435–444
20. Diament BJ, Noble WS (2011) Faster sequest searching for peptide identification from tandem mass spectra. J Proteome Res 10(9):3871–3879. arXiv:http://pubs.acs.org/doi/pdf/10.1021/pr101196n
21. Park CY, Klammer AA, Kall L, MacCoss MJ, Noble WS (2008) Rapid and accurate peptide identification from tandem mass spectra. J Proteome Res 7(7):3022–3027. arXiv:http://pubs.acs.org/doi/pdf/10.1021/pr800127y
22. Käll L, Canterbury JD, Weston J, Noble WS, MacCoss MJ (2007) Semi-supervised learning for peptide identification from shotgun proteomics datasets. Nat Methods 4(11):923

23. Saeed F, Pisitkun T, Hoffert JD, Wang G, Gucek M, Knepper MA (2012) An efficient dynamic programming algorithm for phosphorylation site assignment of large-scale mass spectrometry data. In: 2012 IEEE international conference on bioinformatics and biomedicine workshops (BIBMW), vol. 11, IEEE, BioMed Central Ltd, pp 618–625

Chapter 6
A Easy to Use Generalized Template to Support Development of GPU Algorithms

Fahad Saeed and Muhammad Haseeb

Computational techniques have taken a new meaning for scientific inquiry in biology especially after the introduction of high-throughput experimental techniques. These instruments can produce massive amounts of data that needs to be processed in a scalable fashion to ensure that we can make sense of these data sets from various sources [2, 3]. As expected, Mass Spectrometry (MS) based omics is essential for precision medicine, cancer research, and drug discovery but the scale at which these data sets needs to be processed is massive (tera- to peta-byte levels) [2–4]. We have also shown that proteomics, and meta-proteomics search can taken impractically long times [5, 6]. which can become a major technical hurdle in investigating these systems biology studies. The existing serial algorithms scale very poorly with increasing size of the data sets, and HPC methods are also shown to be much less than optimal [2, 7].

The post-Moore era of computer architectures has given us ubiquitous access to multicore, manycore, CPU-FPGA, and CPU-GPU architectures which can be used for acceleration of applications [8, 9]. However, up until recently there has not been a serious effort toward developing high-performance computing algorithms for MS-based omics. Likewise, any underlying high-performance computing building blocks [8, 9] that are domain specific had not been built to date. However, it is clear that high-performance computing architectures can, and must, be used for accelerating the processing of big MS-based omics data; something that has been successful in so many other domains [10, 11].

One exciting development in computer architecture is the development of Graphics Processing Unit (GPU) which is a low-cost device but is capable of housing thousands of small computational cores which can be used for exploitation of parallelism in many scientific workflows [12, 13]. However, the limitation of GPU-based computing is that the parallel algorithm has to be designed specifically to exploit GPU

Some parts of this chapter may have appeared in [1].

F. Saeed and M. Haseeb, *High-Performance Algorithms for Mass Spectrometry-Based Omics*, Computational Biology 34,
https://doi.org/10.1007/978-3-031-01960-9_6

architecture for accelerating the code. Since the design of the algorithms is application specific, many of the GPU-based algorithms do not have re-usable designs that can be used for MS-based omics. Other limitations include the development of parallel code without taking into account all of the factors that can affect the speedups and scalability (i.e. inserting parallel programming pragmas in code in the hope to get some speedups) leading to plethora of poorly designed parallel algorithm that can exploit the GPU for MS-based omics [8].

When we started this research, we did not just wanted to have another GPU-based algorithms for MS-based omics. We wanted to develop the fundamental building blocks for GPU-based MS omics data analysis that can be used as fundamental guidelines, and generic principles which would give speedups for many workflows. Such generic goalposts would allow MS domain scientists to develop GPU-based algorithm, which would scale, without worrying too much about complexities of GPU architectures (something that may not be possible for domain scientists to pick just because it is a whole other field of study).

To facilitate the development of GPU-based MS omics workflows; we designed, and implemented a generic GPU-based algorithmic design templates called GPU-DAEMON (GPU Algorithm **D**esign, D**a**ta **M**anagement and **O**ptimizatio**n**). GPU-DAEMON is a GPU-based template that allows exploitation of GPU architecture for any data that looks like large number of very small arrays. Of course, it is designed and implemented with MS data analysis in mind, but we showed in our paper that it is very much applicable to other domains (such a fMRI-based neuroscience) analysis. To design GPU-DAEMON we considered all possible bottlenecks, template design for efficient data management of array structures, and various optimization that allows maximal occupancy, and performance for the GPU-based cores. Once we have introduced the design of GPU-DAEMON, we will implement a GPU-based MS-REDUCE algorithm in the next chapter.

6.1 GPU Architecture and CUDA

Graphical processing units (GPU) were developed to improve the graphics (and related) quality for gaming. However, computational scientists have worked toward using this architecture for general purpose computing to improve the scalability of the scientific workflows. GPU consists of large number of cores that can process in parallel, and can be potentially used for processing individual elements of an image matrix, or matrix calculations to exploit the massively parallel cores available in a typical GPU [14]. A typical GPU consists of several *Streaming Multiprocessors (SM)* each of which contains several CUDA cores which can vary from one GPU model to another. For example, GTX 1080Ti GPU contains 28 SMs with 128 CUDA cores each where as K-40 Tesla GPU contains 15 streaming multiprocessors with 192 core each making a total of 2880 cores.

Irrespective of how many SMs or CUDA cores a GPU has; each SM in a GPU has a fast on-chip memory associated with it and shared among its cores. This fast

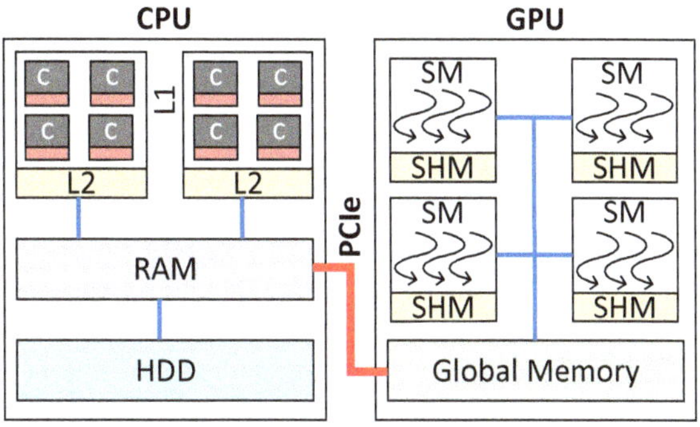

Fig. 6.1 Figure showing CPU-GPU architecture overview. All the data transfers happen via PCIe

on-chip memory is at least 100x times faster than the GPU global memory but is small in size (32kbyte or 64kbyte) [15]. There is also an off-chip memory, known as Global Memory, which is much larger in size (several GB is not uncommon in new GPUs), and is mostly used for storing data, and communicating it with the host CPU. A generic overview of the CPU-GPU architecture is shown in Fig. 6.1.

6.1.1 CUDA Overview

Interest in designing and implementing general purpose computing on GPUs made NVIDIA introduce CUDA standard which can be used with multiple languages to program GPUs [16]. CUDA uses SIMT (Single Instruction Multiple Thread) model which combined the SIMD (Single Instruction Multiple Data) with the assumption of multiple threads and allows exploitation of two levels of parallelism [17, 18]. CUDA standard forms a software overlay that could allow the programmer easy access to parallel architectural features of a GPU. In CUDA programming model, each compute unit is arranged in the form of a Grid of Blocks each containing several threads, where the number of threads, and blocks are GPU dependent. Each thread within a block is assigned 2 IDs, i.e. Threads ID, and Block ID which can be used to track and use each thread in each block. The SMs that are shown in Fig. 6.1 would be replaced with blocks and CUDA cores with threads. As discussed earlier, the number of threads that are active at any given time is dependent on the GPU card that is used [16].

6.1.2 CPU-GPU Computing

For any CPU-GPU computing, one needs a CPU that can act as a host which can offload tasks and data to the GPU which usually behaves like a co-processor. Data from RAM associated with CPU is transferred to the GPU's global memory via

PCIe bus with a set of CUDA instructions. CUDA *kernel* is executed on the GPU where each CUDA core complete the instructions independently. Once the kernel has completed the instructions, and calculation on a given data fragment, the results are transferred back to the host again via PCIe bus. Since PCIe bus is a bottleneck when processing big data, one has to design GPU-based algorithm by having a good understanding (and profiling) of the algorithm [19]. To ensure that scalable processing can happen only most compute intense parts are transferred to the GPU for processing. If enough effort is not invested in the design of the algorithm; the CPU-GPU code will perform poorly when compared to a single one threaded CPU [14]. In the next section we are going to discuss different challenges, and their solutions that one has to consider when design GPU-based algorithms for MS-based omics.

6.2 Challenges in GPU Algorithm Design

This section is dedicated to discuss different challenges, bottleneck, and their solutions when designed GPU-based algorithms for MS-based omics.

6.2.1 Need for Data Parallel Design

GPU compute nodes are generally large in number, and are simple cores without deep pipelines or complex architecture-specific optimizations. Therefore, the best way to exploit GPU cores is to be able to design a parallel strategy that can exploit the data parallelism, and can execute tasks in a parallel fashion without any communication between different cores.

6.2.2 Data Transfer Bottlenecks

Part of the algorithm which is offloaded to the GPU for processing requires that the data is present in the memory of the CPU before the kernel is launched. Since this transfer of data needs to take place via PCIe bus; it is one of the biggest bottlenecks that are faced for big data applications. Needless to say, like any parallel algorithms, if the communication time (to transfer the data) is larger than the time it takes to compute the data; then the scalability of the implementation is limited. In the same way, if the data that is transferred from the GPU to the CPU is larger than the available memory; again efficient data transfer techniques are needed to eliminate or reduce CPU-GPU bottlenecks.

6.2.3 *Non-coalesced Memory Accesses*

GPU threads that are active are grouped into 32 thread-chunks and are known as *warp* which are scheduled on to the SMs when they are available. These wraps can then be mapped to each of the SMs as the resources become available. Global memory accesses from the threads of a wrap can be combined together (colloquially known as coalesced together) to the same memory transaction if the locations have spatial locality . If spatial locality is not present, then accessing the global memory in multiple transactions leads to stalling the wrap execution. This specific issue can become a major bottleneck if the parallel method that can exploit a GPU is not carefully designed.

6.2.4 *Warp Divergence*

In SIMT execution, the thread in a wrap execute in a lock-step which allows all the independent instruction to be executed simultaneously. Branches on the other hand can lead the threads to diverge which can result in efficiency loss, and minimization of this warp-divergence is one of the challenges, and major design decision for parallel algorithms that can exploit GPUs.

6.2.5 *Exploiting Coarse Grained and Fine Grained Parallelism*

Since GPUs require two-levels of parallelism, each level would need fine-grained data management methods to exploit the parallelism that might be available in the architecture. Most often this will require that the data is managed in a way that decomposes the data in fine-grained sets so that they can be processed in parallel. If the decomposition is not fine-grained enough, the amount of parallelism that can be exploited using GPU would be under utilized leading to less scalable solutions.

6.3 Basic Principles of GPU-DAEMON

The proposed GPU base template provides a design that can be used for CPU-GPU-based algorithms for big data omics especially for MS-based omics, Next Generation Sequencing (NGS) based genomics, and fMRI based connectomics. For our purposes, we will focus on the design principles that are relevant to MS-based omics data analysis. The design of our proposed GPU-DAEMON is divided into seven steps where each step gives a generic solution that tackles one or more GPU bottleneck. Of

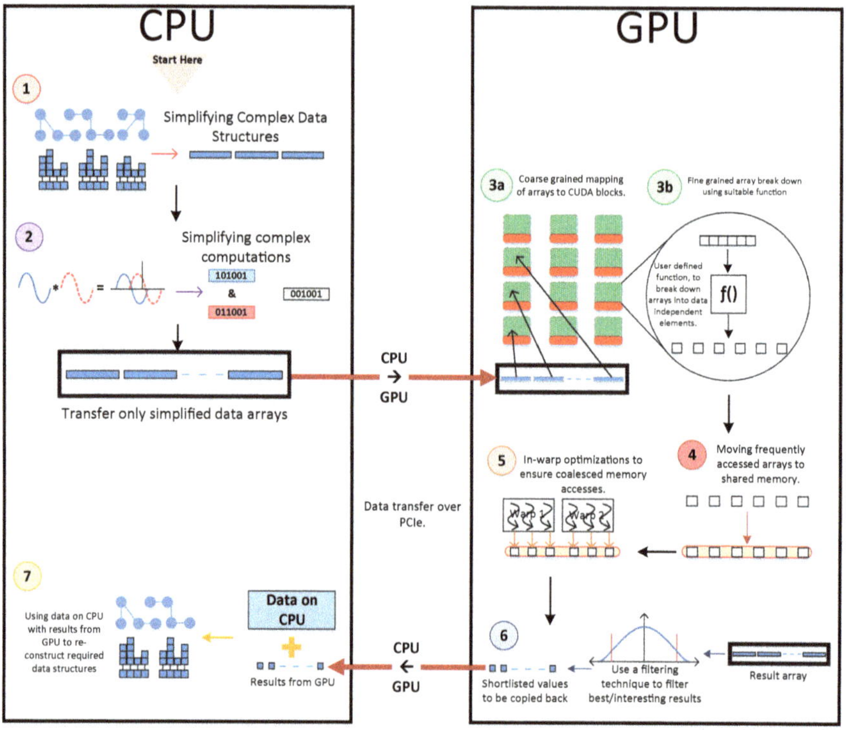

Fig. 6.2 Figure shows the template for GPU-DAEMON

course, these solutions need to be adapted to the application that is being considered but give a good starting point to modify depending on the problem being solved. Figure 6.2 shows the steps in GPU-DAEMON. The first step is to analyze and profile the algorithm for which the GPU-based method is being developed. This step will determine if the proposed method is compute- or data-intensive. Any compute- or data-intensive parts are kept for the GPU-side while other simpler operations may be completed on a CPU.

6.3.1 *Simplifying Complex Data Structures*

PCIe bus that is mostly used to transfer the large data structures is one of the first bottlenecks that are encountered. Our proposed approach of transferring the complete data structures is easy and intuitive for the programmer, and generally does not require the complete redesign of the parallel code. With limited memory compute area in the GPU dictates that only a small portion of the data structure is needed for computation at any given point in time. Therefore, a design methodology that requires one to only transfer part of the data structure will considerably reduce the bottleneck that is associated with PCIe bus.

6.3.2 Simplifying Complex Computations

In general, existing GPU architectures are simple computing platforms and do not allow complex computations which makes them specialized for large number of computations. To this end, the second step of the GPU-DAEMON is to simplify the complex computations, e.g. by converting floating point numbers into integers, or representing those data sets as binary making data computations simpler for these cores. Of course, these simplifications are application specific, and not all computations or data can be made simpler especially in the scenario where precision is required.

6.3.3 Efficient Array Management in GPU

From the data management prospective, CPU-GPU strategy is dependent on how the arrays of the data are managed, and how different CUDA compute nodes compete for access to data; and how this data is accessed can determine the performance of the GPU-based parallel algorithm. To exploit the multiple levels of parallelism that are available, we introduce and evaluate fragmentation strategy that can be accomplished using two steps: (1) First step consists of ensuring that each array can be mapped to a unique block in the GPU in a coarse-grained fashion. This can almost always be accomplished since the number of CUDA blacks are much larger than the number of arrays.

(2) The second step is used to exploit fine-grained parallelism by decomposing each array into sub-arrays and then mapping them to a cluster of threads. Since this is more specific to the application, the data mapping can be categorized into two parts each with a different approach. If the data calculations are independent and are not dependent on the calculation from other arrays cells then an array of size m, following number of elements assigned per-thread should suffice:

$$E_i = \frac{m}{nT} \tag{6.1}$$

$$E_{nT-1} = E_i + m \bmod (nT)$$

where E_i is the number of elements to be mapped to thread i where nT is the total number of threads available per block, and E_{nT-1} represents the number of elements mapped to the last thread in block. Here we assume that Thread IDs start at position 0. Start and end indices for sub-array assigned to each thread can be calculated as

$$SI_i = i * E_i$$

$$EI_i = (SI_i + (i + 1) * E_i) - 1$$

$$EI_{nT-1} = (EI_{nT-2} + E_{nT-1}) - 1$$

Here SI_i and EI_i are the locations for first and last elements of the sub-arrays assigned to thread i, respectively.

When the data is dependent on each other for calculations, then elements need to be divided into data independent subsets using a suitable user defined function as shown in GPU-DAEMON template Fig. 6.2. We denote this function by F_{sub}.

6.3.4 Exploiting Shared Memory

Shared memory that is available on a GPU is 100x times faster than any other memory, which makes exploitation of this memory module most consequential for the performance of the code. The programmer would want to make sure that the frequently accessed part of the calculations is moved to the shared memory. One condition [15, 19] that would ensure that such move will give reasonable speed advantage the following equation must hold:

$$(T_{tf}) + (P_{SM}) < (P_{GM}) \tag{6.2}$$

Here T_{tf} is the time to move data from global to shared memory while P_{SM} and P_{GM} are the processing times in Shared and Global memory, respectively.

6.3.5 In-Warp Optimizations

Optimization strategies which will ensure that we get the best performance from a GPU; we want to ensure that thread divergence inside a warp, and memory coalescing to access global memory are reduced to minimize loss in performance. For thread divergence, usually the parallel algorithm needs to be redesigned (or at least reconfigured) so that the threads needed for computations do not diverge in a warp. Global memory coalescing can be achieved by good thread to data mapping strategy. The mappings discussed in third step of GPU-DAEMON simplify this mapping. By mapping consecutive threads to independent contiguous array segments of Step 3 can help achieve memory coalescing.

6.3.6 Result Sifting

Once computations have been completed, the programmer has to ensure that the output is also managed correctly. Sometimes these output arrays can be larger than the input data [20], and attention has to be paid to ensure that memory transfer bottlenecks are not formed. Usual techniques include either compressing the output results or copying back just the relevant part of the calculations. Since these are very specific to the application under consideration, we will not generalize this for GPU-DAEMON.

6.3.7 Post Processing Results

If in the first step, if a transformation is performed on the data to simplify the transfer and processing then there may be a need for a post-processing phase. This phase is mostly performed on the host processor and is basically an inverse of the data transformation performed in the first step.

6.3.8 Time Complexity Model for GPU-DAEMON

Any algorithm developed using GPU-DAEMON will have total time T_{tot} comprising of two terms

$$T_{tot} = T_{CPU} + T_{GPU}$$

where T_{CPU} is the total time complexity of CPU part of the design and T_{GPU} is the total time complexity of GPU part of the design. Here we will give a generic formulation for T_{GPU}, this formulation can be used to derive the actual time complexity of the GPU part of the algorithm. T_{GPU} depends on the time taken to disintegrate a given array into data independent segments (T_{sub}), time for processing the data independent arrays (T_{proc}) and the time for result sifting step (T_{sift}), i.e. $T_{GPU} = T_{sub} + T_{proc} + T_{sift}$. If we consider N arrays with each of size n then the total time for applying disintegration function f_{sub} to N arrays on GPU would be equal to $T_{sub} = \frac{N}{B} * (\frac{T(f_{sub})}{p})$ where B is the number of Cuda Blocks active at a given time, p is the number of threads active per block and $T(f_{sub})$ is the time for f_{sub}. Similarly, we can compute $T(f_{proc})$ to be $\frac{N}{B} * (\frac{T(f_{proc})}{p})$ for processing function f_{sub} and $T_{sift} = \frac{N*x*T(f_{sift})}{B*p}$ for result sifting function f_{sift}. Here x is the number of elements in each result array. This gives us

$$T_{GPU} = \frac{N}{B * p} * (T(f_{sub}) + T(f_{proc}) + x * T(f_{sift})) \tag{6.3}$$

References

1. Awan MG, Eslami T, Saeed F (2018) Gpu-daemon: Gpu algorithm design, data management and optimization template for array based big omics data. Comput Biol Med 101:163–173
2. Abuín JM, Pichel JC, Pena TF, Amigo J (2016) Sparkbwa: speeding up the alignment of high-throughput dna sequencing data. PloS one 11(5):e0155461
3. Awan MG, Saeed F (2016) Ms-reduce: an ultrafast technique for reduction of big mass spectrometry data for high-throughput processing. Bioinformatics 32(10):1518–1526
4. Saeed F, Hoffert JD, Knepper MA (2013) Cams-rs: clustering algorithm for large-scale mass spectrometry data using restricted search space and intelligent random sampling. IEEE/ACM Trans Comput Biol Bioinform 11(1):128–141
5. Kong AT, Leprevost FV, Avtonomov DM, Mellacheruvu D, Nesvizhskii AI (2017) Msfragger: ultrafast and comprehensive peptide identification in mass spectrometry-based proteomics. Nat Methods 14(5):513–520
6. Jagtap P, Goslinga J, Kooren JA, McGowan T, Wroblewski MS, Seymour SL, Griffin TJ (2013) A two-step database search method improves sensitivity in peptide sequence matches for metaproteomics and proteogenomics studies. Proteomics 13(8):1352–1357
7. Saeed F (2015) Big data proteogenomics and high performance computing: challenges and opportunities. In: 2015 IEEE global conference on signal and information processing (GlobalSIP). IEEE, pp 141–145
8. Tariq U, Cheema UI, Saeed F (2017) Power-efficient and highly scalable parallel graph sampling using fpgas. In: 2017 international conference on ReConFigurable computing and FPGAs (ReConFig). IEEE, pp 1–6
9. Eslami T, Awan MG, Saeed F (2017) Gpu-pcc: a gpu based technique to compute pairwise pearson's correlation coefficients for big fmri data. In: Proceedings of the 8th ACM international conference on bioinformatics, computational biology, and health informatics. ACM, pp 723–728
10. Lin C-H, Li J-C, Liu C-H, Chang S-C (2017) Perfect hashing based parallel algorithms for multiple string matching on graphic processing units. In: IEEE transactions on parallel and distributed systems
11. Ma Y, Chen L, Liu P, Lu K (2016) Parallel programing templates for remote sensing image processing on gpu architectures: design and implementation. Computing 98(1–2):7–33
12. Warris S, Yalcin F, Jackson KJ, Nap JP (2015) Flexible, fast and accurate sequence alignment profiling on gpgpu with paswas. PloS one 10(4):e0122524
13. Baumgardner LA, Shanmugam AK, Lam H, Eng JK, Martin DB (2011) Fast parallel tandem mass spectral library searching using gpu hardware acceleration. J Proteome Res 10(6):2882–2888
14. Fatahalian K, Sugerman J, Hanrahan P (2004) Understanding the efficiency of gpu algorithms for matrix-matrix multiplication. In: Proceedings of the ACM SIGGRAPH/EUROGRAPHICS conference on graphics hardware. ACM, pp 133–137
15. Awan MG, Saeed F (2016) Gpu-arraysort: a parallel, in-place algorithm for sorting large number of arrays. In: 2016 45th International Conference on Parallel Processing Parallel Processing Workshops (ICPPW). IEEE, pp 78–87
16. Nvidia (2016). http://docs.nvidia.com/cuda/index.htmlCUDA Toolkit Documentation v7.5. http://docs.nvidia.com/cuda/index.html
17. Nickolls J, Buck I, Garland M, Skadron K (2008) Scalable parallel programming with cuda. Queue 6(2):40–53
18. Lindholm E, Nickolls J, Oberman S, Montrym J (2008) Nvidia tesla: a unified graphics and computing architecture. IEEE Micro 28(2)

19. Awan MG, Saeed F (2017) An out-of-core gpu based dimensionality reduction algorithm for big mass spectrometry data and its application in bottom-up proteomics. In: Proceedings of the 8th ACM international conference on bioinformatics, computational biology, and health informatics. ACM, pp 550–555
20. Lee J-Y, Fujimoto GM, Wilson R, Wiley HS, Payne SH (2017) Blazing signature filter: a library for fast pairwise similarity comparisons. bioRxiv 162750

Chapter 7
Computational CPU-GPU Template for Pre-processing of Floating-Point MS Data

Fahad Saeed and Muhammad Haseeb

The data from MS spectra is usually stored as shortlist of numbers. To process these spectra, more often than not, one has to "see" inside the data to make data pre- and post-processing decisions [2]. Sorting, and searching of data for an array of numbers is one of the oldest problems in computer science. There has been significant effort in developing algorithms that can sort very large array [3]. However, for MS data, instead of having a single large array (of m/z values) there are a lot of moderately sized arrays of a very large number. We have demonstrated that sorting (or searching) of MS data with a lot of spectra is a bottleneck for many pre-processing routines [2, 4]. In order to make this pre-processing efficient, and allowing the users to be able to use our proposed techniques we have formulated a template-based GPU strategy known as GPU-DAEMON. GPU-DAEMON is a strategy that allows developers who might not be familiar with CPU-GPU architecture but would want to utilize the parallel strategy for efficient processing. We have developed a GPU-Array sort algorithm that first appeared in our paper [4] which allows us to utilize CPU-GPU architecture, and sort millions of short MS spectra. Figure 7.1 shows design of GPU-ArraySort overlaid on GPU-DAEMON template.

7.1 Simplifying Complex Data Structures

Since GPU-ArraySort is mostly used as integral part of a bigger algorithm, in this step the data to be sorted can be extracted from larger data structures and stored in the form of simple arrays. These arrays are then transferred over to GPU memory via the PCIe cable. Since the sorting operation cannot be further simplified, the step for simplification of computations was skipped for this algorithm.

Some parts of this chapter may have appeared in [1].

F. Saeed and M. Haseeb, *High-Performance Algorithms for Mass Spectrometry-Based Omics*, Computational Biology 34,
https://doi.org/10.1007/978-3-031-01960-9_7

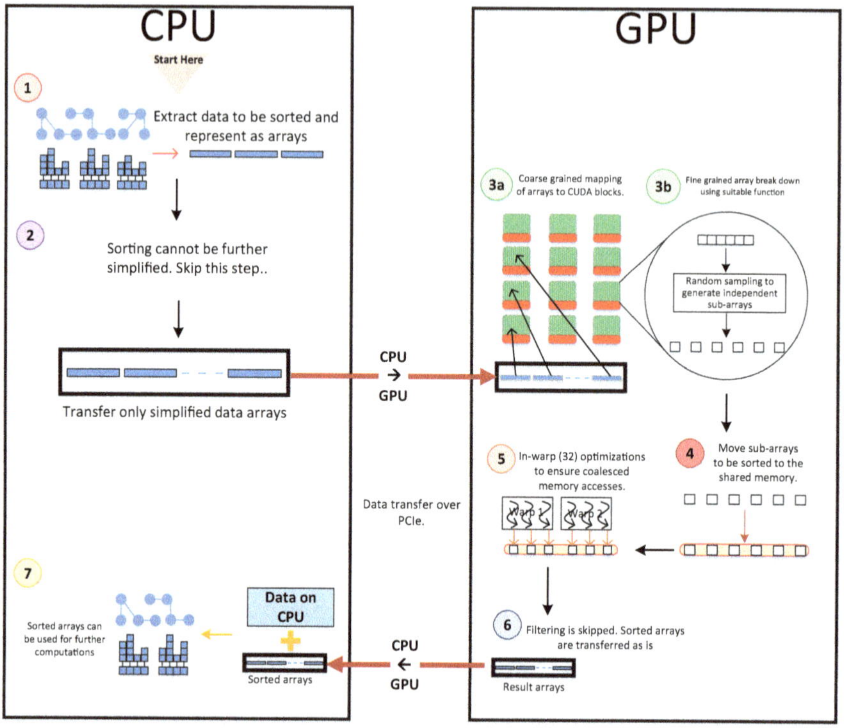

Fig. 7.1 Design of GPU-ArraySort overlaid on GPU-DAEMON template

7.2 Efficient Array Management

Sorting problem allows placement of the elements depending on the stored value which is presented as dependent sub-array case of Sect. 6.3.3. We get the coarse-grained mapping of each array on CUDA blocks as discussed in Sect. 6.3.3. We achieve coarse-grained mapping of each array on different CUDA blocks using the method discussed in Sect. 6.3.3. Sample-based bucketing technique was performed to exploit fine-grained parallelism [5]. We used this strategy to fragment data into sub-arrays which are mapped on the compute units. Sample-based bucked functioning (F_{sub}) for splitter and bucketing is discussed in the section below.

7.2.1 Splitter Selection

The arrays are assumed to be small enough that it fits within GPU's shared memory, and the number of splitters required depends on the number of buckets needed for the array. The size of these buckets must be optimized to get maximum efficiency.

Our empirical study [1] has shown that the best performance is gained when the number of elements is no more than 20 [1]. This choice of size of bucket is totally independent of size of individual array as well as total number of arrays.

Definition 7.1 If n is the size of an array, let B_i be the set of buckets for array i, $B_i = \{b_1, b_2, b_3, \ldots b_p\}$ where $p = \lfloor \frac{n}{20} \rfloor$.

For p buckets we need to have $p - 1$ splitters, these splitters are obtained from a sample set obtained from the unsorted array A_i using regular sampling method. Our studies [1, 6] have shown that if the data is uniformly distributed then 10% regular sampling resulted in the most load-balanced buckets. The samples are obtained first using the in-place insertion sort, and then the remaining $p - 1$ splitters are chosen by picking splitter at *regular* intervals. Each block then returns to its splitter that is written to the global memory at indices calculated using the block ids that are written in consecutive memory locations with each block performing all operations using a single thread. Since the sampled array is small it can be placed inside the memory and the proposed technique is efficient to be used.

The array of splitters thus formed can be defined as

Definition 7.2 Let S be the array of size N, each element $s_i \in S$ is an array of size q which consists of splitters for array A_i. $S = \{s_1, s_2, s_3, \ldots, s_N\}$ where $s_i = \{sp_1, sp_2, sp_3 \ldots, sp_q\}$ and $q = p - 1$.

Algorithm 5 describes a per-thread pseudo-code for first phase.

7.2.2 *Bucketing*

In this phase, the splitter values obtained from the previous phase are translated into a global array used to keep track of the bucket sizes.

Definition 7.3 Let Z be the array of size N, each element $z_i \in Z$ is an array of size q which consists of bucket sizes for array A_i. $Z = \{z_1, z_2, z_3, \ldots, z_N\}$ where $z_i = \{zb_1, zb_2, zb_3 \ldots, zb_p\}$, here each $zb_j \in z_i$ represents size of bucket j in array A_i.

Each array is assigned a unique block with threads equal to the number of buckets p. The sub-array sp_i is small in size and can be moved to the shared memory for effective and frequent usage. The pointer to these arrays can also be determined on the fly depending on the block and thread id such that each thread gets a unique pair of splitters.

Definition 7.4 Let r_i denote a splitter pair for a thread i then $r_i = \{sp_i[tid], sp_i[tid + 1]\}$ here tid denotes each thread's id. The splitter pair allows us to have thread that avoids branch divergence by removing other paths of the code as observed in Algorithm 4. To avoid any overlapping buckets, we use two additional splitters in sub-array sp_i by adding a splitter smaller than the smallest, and larger than the largest value in A_i. Keeping track of a counter $zb_j \in z_i$, where j is the bucket and i is the array, array A_i can be traversed in parallel to complete the bucketing process which will result in each counter containing the size of the bucket. Each bucket is written back to the actual memory location of array A_i. Using this method we are able to parallelize this write back process with the advantage of saving more than 50% of device's global memory.

Algorithm 4 describes a per-thread pseudo code for second phase.

Algorithm 4: Per thread pseudo code for bucketing phase

Data: An array A_i and a pair of splitters r_i
Result: A bucket of elements within splitter pair range

splitterPair = obtainSplitters(r_i)
initializeBucket(bucket)
index = 0
bucketIndex = 0
while *not the end of array* A_i **do**
  if *splitterPair[1]* < A_i*[index]* < *splitterPair[2]* **then**
    bucket[bucketIndex] = A_i[index] bucketIndex ++
  index ++

7.3 In-Wrap Optimizations and Exploiting Shared Memory

The buckets and the sub-arrays, formed in the previous step, are small enough to fit in the shared memory. However, the sub-arrays are assigned to a warp which are placed in contiguous location in the memory to minimize memory transactions and accesses. If the size of the array is larger than the GPU memory; it must be sorted in batches. Our design and results have shown that CUDA streams data transfer, and data processing times overlap to create a pipeline like affect resulting in better processing times as compared to simple batch processing.

7.4 Time Complexity Model

Time complexity of GPU-ArraySort can be determined by replacing the values of $T(f_{sub})$ and $T(f_{proc})$ in Eq. 6.3 with $O(\frac{n}{p})$ and $O(\frac{n}{p} * log(\frac{n}{p}))$, respectively. Here n is the length of each array while p is the number of threads per CUDA block.

$$O\left(\frac{n}{p} + \frac{n}{p} * log\left(\frac{n}{p}\right)\right) \tag{7.1}$$

Algorithm 5: Per thread pseudo code for splitter selection

Data: An array A_i and required number of splitters q
Result: An array of splitters s_i for array A_i

```
samples = obtainSamples(A_i)
sortedSamples = insertionSort(samples)
index = 0
sampleIndex = 0
stride = calculateStride(sortedSamples)
while sizeOf(s_i) not equal to q do
    s_i[index] = A_i[sampleIndex]
    sampleIndex+ = stride
    index + +
```

7.5 Performance Evaluation

Performance evaluation of the proposed technique was performed by sorting large number of arrays, and then see which techniques can have a higher throughput (i.e. can sort more number of arrays on a given GPU). Since there is no dedicated GPU-based algorithm we used NVIDIA's Thrust Library to used stable sort by tagging them with keys. A brief explanation is given below:

7.5.1 *Sorting Using Tagged Approach (STA)*

Let $I = \{A_1, A_2, A_3, \ldots, A_i\}$ be a list of arrays to be sorted where $i = N$, then in order to use the STA approach we create another list of arrays and call it the array of tags.

Definition 7.5 Let $T = \{T_1, T_2, T_3, \ldots, T_i\}$ be list of arrays of tags such that $i = N$ and $|T_i| = |A_i|$. Here each element $t \in T_i$ represents a tag for array T_i and carries the same value, i.e. $t = i$. Once the tags have been created all the arrays of I are merged into one single array and all the tags are merged into another array. Then the sorting proceeds in two steps :

- Perform a stable sort on the array, containing the arrays to be sorted, using the array of tags, as keys.
- Perform a stable sort on the array of tags, using the array of arrays to be sorted, as keys.

The process has been explained in Fig. 7.2. It is clear that STA kind of strategy takes a lot more resources (both memory, and time) than would be needed for an optimal parallel computing strategy. Redundant work includes adding tags to the array, sorting them, and the need to sort the tag arrays in the GPU global memory. Further the STA technique uses Radix sort for sorting of the numbers which utilizes

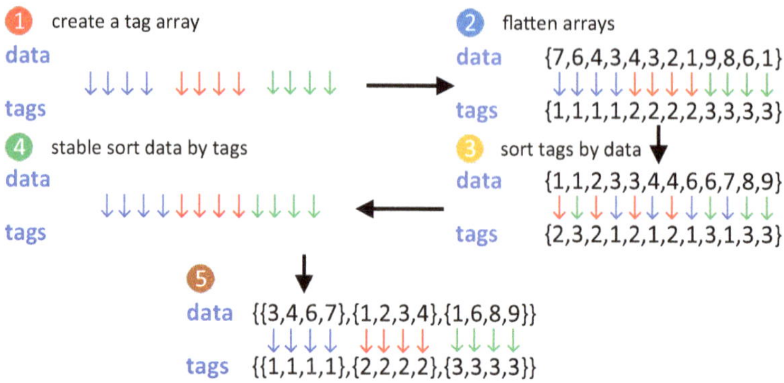

Fig. 7.2 A step by step process explaining the STA technique, here the arrays to be sorted are referred as test arrays: (I) A tag array is created for each array to be sorted. (II) Arrays are merged into one big array. (III) Arrays are sorted using the array of tags as keys. (IV) Again arrays are sorted using the test arrays as key. (V) Arrays are restored based upon their tags

almost $O(N)$ more space than the data under process [7]. The estimated memory that is used by STA is approx. 3 times more than the memory required for optimal processing [8] be required to sort all the arrays.

7.5.2 *Runtime Analysis and Comparisons*

We performed the experiments to create 4 different data sets where each set consists of 200k arrays where each array was generated using uniform distribution between 0 and $2^{31} - 1$. The size of these arrays was 1000, 2000, 3000, and 4000 respectively for four data sets that were produced with floating point data type. All the experiments listed below were performed using 24 CPU cores each operating at 1200 MHz, Graphic Processing Unit used was NVIDIA's Tesla K-40c consisting of 2880 CUDA cores. Total global memory available on the device was 11520 MBytes and the shared memory of 48 KBytes was available per block.

Figures 7.3, 7.4, 7.5 and 7.6 show runtime comparison between STA and GPU-ArraySort. GPU-Array sort with its superior memory efficient design out-performs STA technique for all the array sizes that we investigated.

7.5.3 *Data Handling Efficiency*

The experiments that we discussed in the previous section were performed again without any bounds on the number of arrays. Table 7.1 shows that the maximum

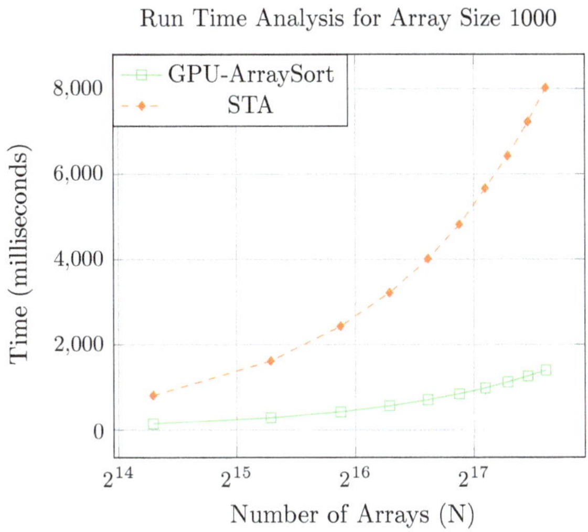

Fig. 7.3 The figure shows time versus number of arrays plots for GPU-ArraySort and the tagged sorting approach using key-based stable sorting algorithm from Thrust library

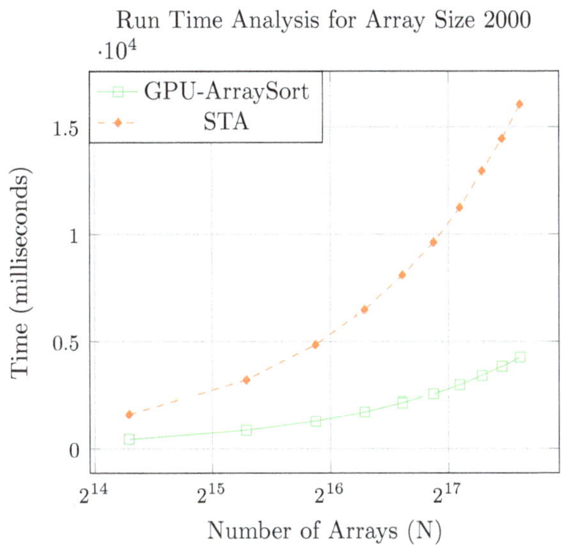

Fig. 7.4 The figure shows time versus number of arrays plots for GPU-ArraySort and the tagged sorting approach using key-based stable sorting algorithm from Thrust library

number of arrays processed by each competing method. These experiments demonstrate that GPU-ArraySort algorithm is able to process more than 3 times more arrays than the competing STA-based approach.

Fig. 7.5 The figure shows time versus number of arrays plots for GPU-ArraySort and the tagged sorting approach using key-based stable sorting algorithm from Thrust library

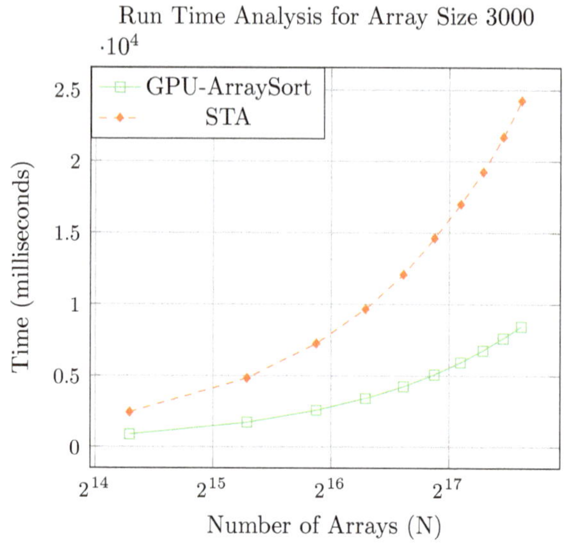

Fig. 7.6 The figure shows time versus number of arrays plots for GPU-ArraySort and the tagged sorting approach using key-based stable sorting algorithm from Thrust library

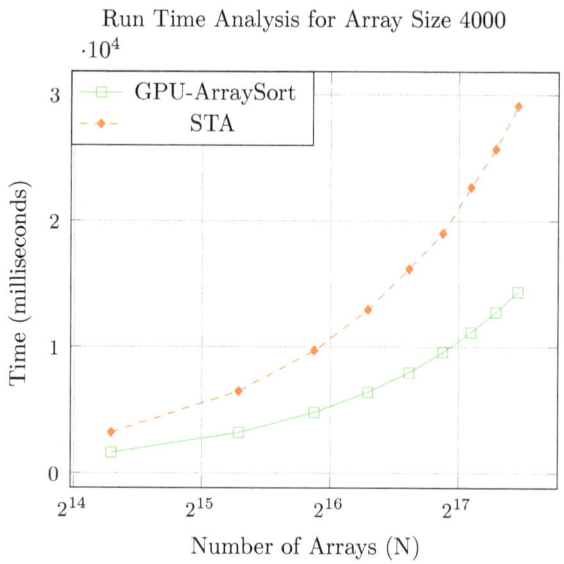

Table 7.1 Table for memory efficiency of GPU-ArraySort

Array size	GPU-ArraySort	STA
1000	2000000	700000
2000	1050000	350000
3000	700000	200000
4000	500000	150000

Note: The table shows number of arrays sorted by STA technique and GPU-ArraySort. The center column shows that GPU-ArraySort can sort upto 2 million arrays of size 1000 while in comparison STA technique was able to sort only 0.7 million arrays. This comparison is for Tesla K-40c GPU

References

1. Awan MG, Saeed F (2016) Gpu-arraysort: a parallel, in-place algorithm for sorting large number of arrays. In: 2016 45th international conference on parallel processing workshops (ICPPW). IEEE, pp 78–87
2. Awan MG, Saeed F (2016) Ms-reduce: an ultrafast technique for reduction of big mass spectrometry data for high-throughput processing. Bioinformatics 32(10):1518–1526
3. Satish N, Harris M, Garland M (2009) Designing efficient sorting algorithms for manycore gpus. In: IEEE international symposium on parallel & distributed processing, 2009. IPDPS 2009. IEEE, pp 1–10
4. Awan MG, Saeed F (2016) Gpu-arraysort: A parallel, in-place algorithm for sorting large number of arrays. In: 2016 45th international conference on parallel processing, parallel processing workshops (ICPPW). IEEE, pp 78–87
5. Liu F, Huang M-C, Liu X-H, Wu E-H (2009) Efficient depth peeling via bucket sort. In: Proceedings of the conference on high performance graphics 2009. ACM, pp 51–57
6. Awan MG, Saeed F (2017) An out-of-core gpu based dimensionality reduction algorithm for big mass spectrometry data and its application in bottom-up proteomics. In: Proceedings of the 8th ACM international conference on bioinformatics, computational biology, and health informatics. ACM, pp 550–555
7. Horsmalahti P (2012) Comparison of bucket sort and radix sort. arXiv:1206.3511
8. Guo Z, Huang T-W, Lin Y (2020) Gpu-accelerated static timing analysis. In: Proceedings of the 39th international conference on computer-aided design, pp 1–9

Chapter 8
G-MSR: A GPU-Based Dimensionality Reduction Algorithm

Fahad Saeed and Muhammad Haseeb

In our previous chapters, we have introduced a generalized strategy has been devised that can be used for processing of MS-based omics data sets on a CPU-GPU architecture. Pre-processing of this data is an essential element for proteomics pipelines but the scalability of these pre-processing workflows has not been the focus of research in this domain. Hence many of the existing pipelines may take multiple hours or days to complete the processing [1].

The number of spectra produced by a single experiment can vary from few thousand to billions depending on the objective of the experiment, and species that are being considered. However, this is just for single run of experiments. The plethora of experimental spectra now available from different laboratories facilitates an enormous amount of data that can be used, reused, or reevaluated for systems biology researchers. Each spectrum consists of 2 columns of data where the first column consists of mass-to-charge ratio (m/z), and the second column consists of the corresponding intensities [2, 3]. In this chapter, we showcase how a generalized GPU-DAEMON strategy can be used for a noise reduction workflow for MS-based omics data using CPU-GPU architectures. Our strategy is called G-MSR and was first introduced in [3]. Note that this GPU-based noise reduction algorithm tries closely follows the processing patterns of MS-REDUCE algorithm [4].[1]

8.1 G-MSR Algorithm

Similar to the MS-REDUCE algorithm, G-MSR need to perform three steps: (1) Spectral classification, (2) Quantization, and (3) Weighted random sampling. The input parameters consist of a reduction factor R which is then applied to the given spectra s which will result in a spectrum that is equal to $R * |s|$. The classification stage allows the spectra to be classified into four classes. The spectra which are

[1] Some parts of this chapter may have appeared in [3].

F. Saeed and M. Haseeb, *High-Performance Algorithms for Mass Spectrometry-Based Omics*, Computational Biology 34,
https://doi.org/10.1007/978-3-031-01960-9_8

classified in the same class are quantized such that each spectrum is grouped based on the peaks that are significant. Then there is a weighted random sampling in used in each class such that peaks are randomly sampled from each quantum. The weighted sampling allows that the most significant peaks make it to the final reduced spectra.

The weighted random sampling step can be formulated as the following equation where x_i is the sampling weight for the ith quantum, q_i is the number of peaks in the quanta i, p' is the total peaks in the spectra, and n is the number of quanta.

$$\sum_{i=0}^{n} \frac{x_i}{100} = p' \tag{8.1}$$

Figure 8.2 shows the design of G-MSR overlapped on the GPU-DAEMON template and Fig. 8.1 shows a comparison in the workflows of MS-REDUCE and G-MSR.

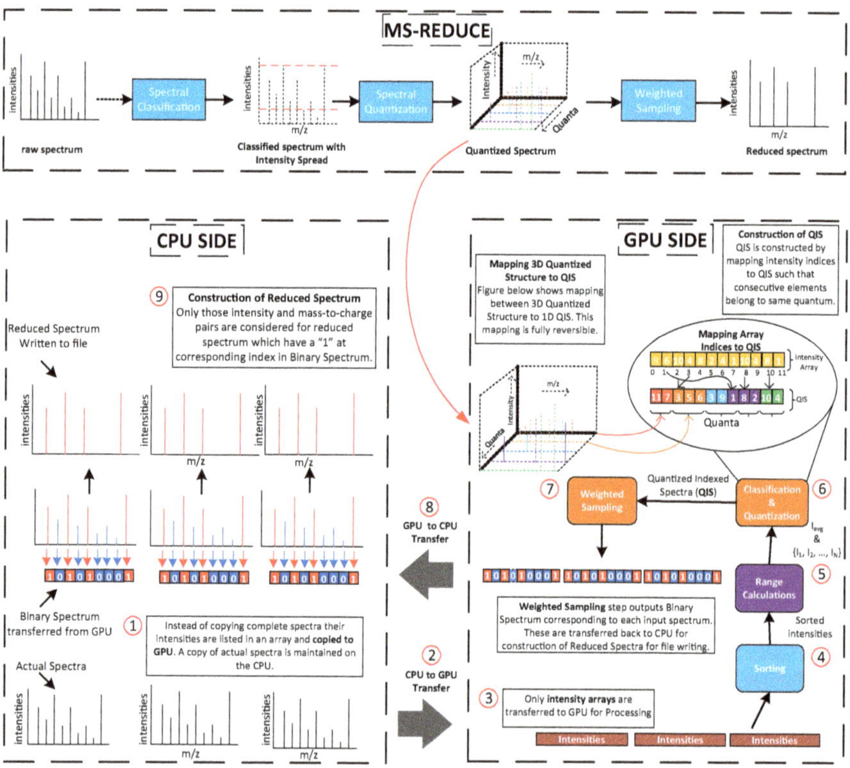

Fig. 8.1 **a** Shows the work flow of MS-REDUCE. **b** Construction of QIS from 3-D quantized spectrum from MS-REDUCE. **c** Work flow of G-MSR, blocks with same color represent processing in same kernel. A copy of actual spectra is maintained on the CPU for the construction of reduced spectra

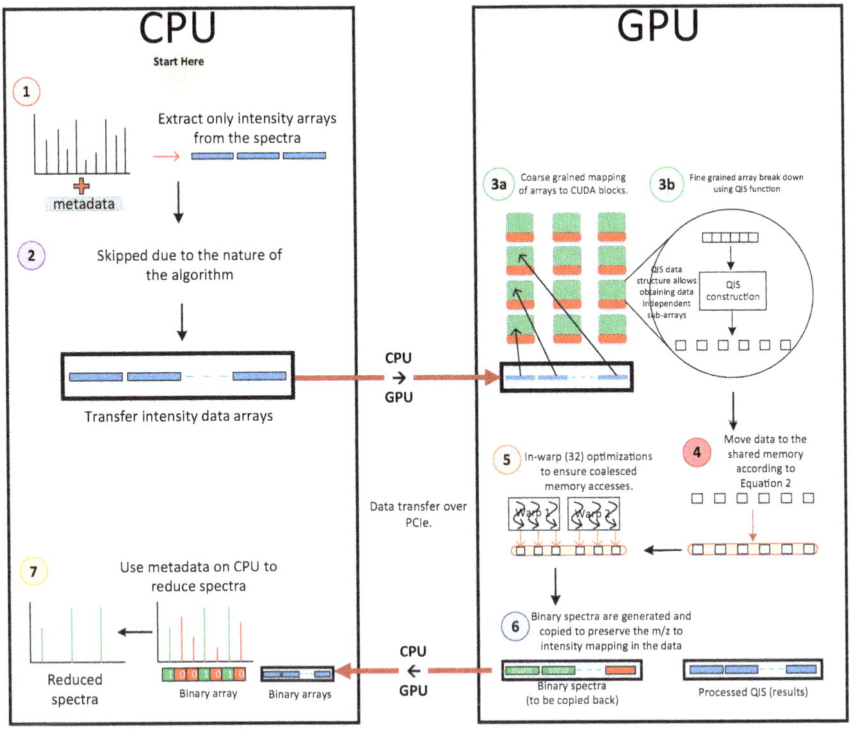

Fig. 8.2 Design of G-MSR overlaid on GPU-DAEMON template

8.1.1 Simplifying Complex Data Structures

As discussed before, the mass spectra obtained from MS consist of mass-to-charge ratios and their corresponding intensities. In a naive method complete spectra along with their meta-data would be transferred over the PCIe cable to GPU for processing. But following the GPU-DAEMON template we separate the intensities from the larger data structure in the forms of multiple arrays (one array for each spectrum) and only transfer these over to GPU memory. This cuts down the amount of data being transferred by more than 50%. The actual spectra are kept on the host for book-keeping and post-processing phase.

8.1.2 Simplifying Complex Computations

Since intensities are floating point numbers, we round them off to nearest integer before transferring them to GPU. This converts all the floating point computations to integer computations thus simplifying the computations. As shown at the end of this chapter, this approximation does not affect the algorithm's performance.

8.1.3 Efficient Array Management

The quantization stage of MS-REDUCE reduction algorithm discussed in Chap. 3 transforms the spectra into 3-Dimensional data structures. Managing this 3-D data structure is challenging for data processing on a GPU architecture [5], also in-order for GPU-DAEMON's array management technique to work we need to map the data into a 1-Dimensional array. To achieve this, we introduced a novel data structure called Quantized Index Spectrum (QIS) which maps a 3-D quantized spectrum onto a 1-D array which can then be easily managed using the techniques discussed in Sect. 6.3.3. The QIS data structure serves a dual purpose of transforming 3-D quantized spectra to 1-D array while performing the step of quantization. As discussed before, the quantization step basically groups together the peaks of a spectrum. In a QIS data structure, these groups of peaks are present in contiguous memory locations, with a separate array of pointers keeping track of starting and ending points. Each of this group can be considered as a sub-array, since these sub-arrays are independent of each other we can use the strategy of Sect. 6.3.3 for exploiting fine-grained parallelism. For G-MSR algorithm, we replace F_{sub} by QIS construction in GPU-DAEMON template. In order to construct a QIS, instead of clustering peaks together as in MS-REDUCE, we clustered together with the indices of the peaks which make up a quantum. For each spectrum, a QIS is an array containing peak indices clustered together at computed distances. We refer to this structure as quantized-indexed-spectrum (QIS). We can formally define QIS for a spectrum s_i as: *Definition:* Q_i where $Q_i = \{q_1, q_2, q_3, \ldots, q_m\}$ and each $q_t = \{l_1, l_2, l_3, \ldots, l_n\}$ is quantum t, and l represents index for a peak in s_i. In QIS structure, quanta are sorted in their increasing order. Figure 8.1b shows the construction of QIS from intensity array. The QIS then overwrites the spectrum to conserve space.

8.1.4 Exploiting Shared Memory

To better exploit the shared memory, sub-arrays are then moved to the shared memory for further processing if the Eq. 6.2 is satisfied.

8.1.5 In-Warp Optimizations

The sub-arrays created by the QIS are a part of a larger array, with their beginning and end pointers listed separately. So, all the sub-arrays created by QIS are in contiguous memory locations. This feature of QIS helps ensure that when consecutive sub-arrays are processed by consecutive threads of a warp, memory coalescing takes place.

8.1.6 Result Sifting

In the first step, rather than transferring complete spectra we transferred only the part which was needed for GPU-processing, and because of the random sampling which takes place in the third phase of dimensionality reduction algorithm [4], it becomes difficult to maintain which intensities are eliminated on the GPU-side. To tackle this problem, we used an additional property of QIS data structure, i.e. the indices of peaks which are eliminated on the GPU-side are retained with a place-holder. These place-holders help in constructing a binary spectrum indicating the indices of intensities to be retained in the reduced spectrum. We define Binary Spectra as *Definition:* Given a spectrum $s_i = \{p_1, p_2, p_3, \ldots, p_n\}$ a Binary Spectrum B_i for the corresponding reduced spectrum s_i' is defined as, $B_i = \{e_j = 1 | p_j \in s_i'\} \cup \{e_j = 0 | p_j \notin s_i'\}$. In other words, if a peak at index j in s_i is included in the reduced spectrum then there will be a 1 at index j of B_i; otherwise it will be zero. For each spectrum, a binary spectrum is generated and only these binary spectra are then copied back to CPU. Binary spectra are memory efficient and helpful in quick reconstruction of reduced spectra on the CPU side. Introduction of QIS and Binary Spectra thus enabled G-MSR to copy back just the bare minimum and resolve the GPU-CPU bottleneck.

8.1.7 Post Processing Results

The Binary Spectra copied back in the previous phase are then used for constructing the reduced spectra on the CPU side as shown in Fig. 8.2. Figure 8.3 shows the difference in the amount of data handled by MS-REDUCE and G-MSR.

8.2 Results and Experiments

8.2.1 Time Complexity Model

To compute the time complexity of G-MSR we replace $T(f_{sub}) = O(\frac{N}{B}) + O(\frac{N*n^2}{B*p}) + O(\frac{N*n}{B})$ and $T(f_{proc}) = O(\frac{s*N}{B})$ in Eq. 6.3. Here the f_{sub} time includes sorting, classification and construction of QIS data structure while the f_{proc} time consists of weighted random sampling phase. Replacing the values in Eq. 6.3 and simplifying leaves us with:

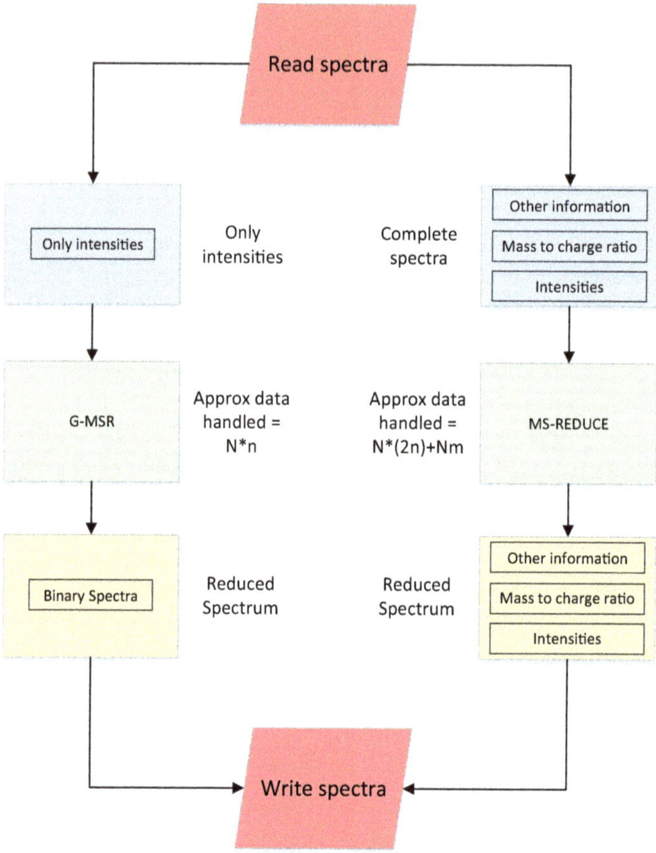

Fig. 8.3 Here N denotes the number of spectra, n the size of largest spectrum, and m the size of other information. Transferring only intensities to GPU for processing can conserve more than 50% of scarce in-core memory

$$O\left(\frac{N*(n^2+l)}{B*p}\right) \tag{8.2}$$

where $l = p*(2+n+n*s)$ and s is the sampling rate.

8.2.2 *Experiment Setup*

For all the experiments we made use of a Linux server running Ubuntu Operating System, version 14.01. The server houses two Intel Xeon E5-2620 Processors, clocked at 2.40 GHz with a total RAM of 48 GBs. The system has an NVIDIA Tesla

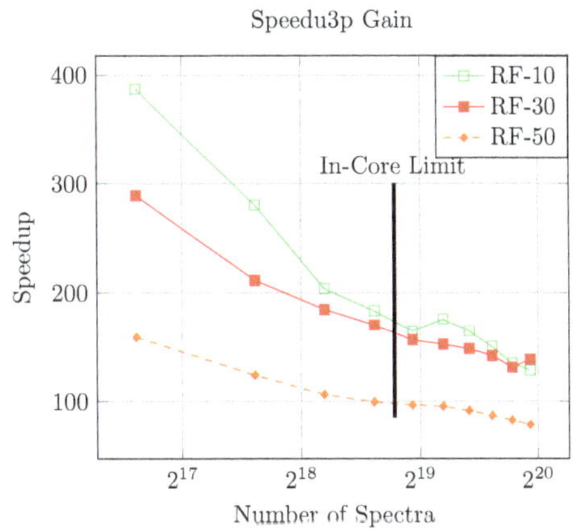

Fig. 8.4 Figure showing the speed up gained by G-MSR over MS-REDUCE while operating at reduction factor (RF) of 10, 30, and 50. The vertical line represents the point where in-core memory is filled

K-40c GPU with a total of 2880 CUDA Cores and 12 GBs of RAM. CUDA version 7.5 and GCC version 4.8.4 were used for compilation.

8.2.3 Scalability and Time Analysis

For this experiment, we used the appended UPS2 dataset which had over a million spectra. Timing experiments were performed with progressively increasing datasizes to cover the cases where data fits in the GPU's memory and when it doesn't. Our experiments showed peak speedups of 386, 288, and 158 for the three Reduction Factors (RF) of 10%, 30%, and 50%, respectively. In accordance with the Eq. 8.2 we get smaller speedup for larger RF and we observe a decrease in speedup with increasing number of spectra in Fig. 8.4. Also with higher RF, amount of data being processed is increased. This increased data leads to more memory being used per warp and thus minimizes the number of concurrent threads leading to increased execution time shown in Fig. 8.5.

8.2.4 Quality Assessment

We used the same method of quality assessment as shown in Fig. 5.6. For our experiments we set the FDR value of interest to 5%, i.e. any PSM having FDR value below 5% is an acceptable match, we call them effective matches. Figure 8.7 shows percentage of effective matches with varying reduction factors for both algorithms. G-MSR and MS-REDUCE gave almost same percentages of effective matches.

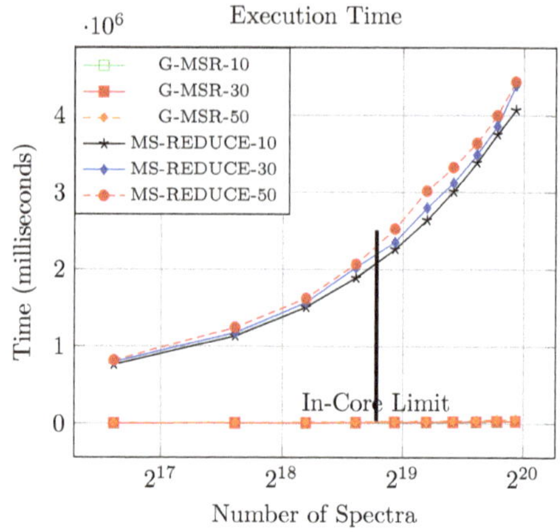

Fig. 8.5 Figure showing the execution times for G-MSR and MS-REDUCE for varying reduction factors. In the legend, numbers following the algorithm names are reduction factors. The vertical line represents the point where in-core memory is filled

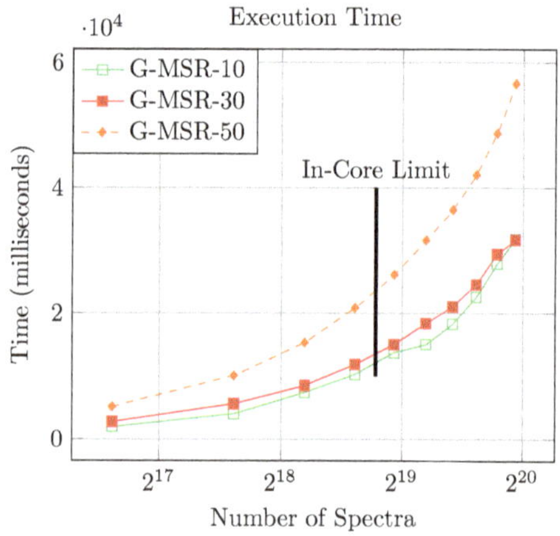

Fig. 8.6 Figure showing the execution times for G-MSR operating at reduction factors of 10, 30, and 50. The vertical line represents the point where in-core memory is filled

8.2.5 Reductive Proteomics for high-resolution instruments

In high-resolution proteomics, the x-axis resolution. i.e. number of bins can lead to large data processing times. Pre-processing of spectra with G-MSR and MS-REDUCE will reduce the dataset size and hence the processing times. We performed peptide deductions for UPS2 dataset after preprocessing it with G-MSR at different reduction factors (RF). We used Tide integrated with hiXcorr was used for peptide deduction in this experiment. Fig. 8.8 shows that the performance of peptide deduction becomes more scalable with smaller reduction factors even with increasing resolution.

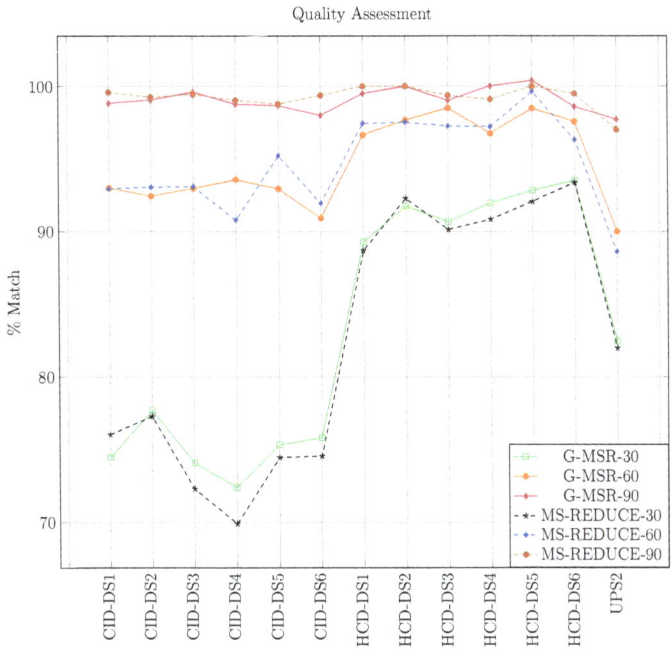

Fig. 8.7 A comparison of quality assessment plots of MS-REDUCE and G-MSR. In the legend, numerical value following name of the algorithm represents its reduction factor. X-axis contains the labels for the experimental datasets while Y-axis represents the percentage of peptide matches with FDR of greater than 5%

8.2.6 Comparison with Unified Memory

To assess the performance of G-MSR (a GPU DAEMON-based version of MS-REDUCE algorithm), we compared it against a unified memory-based GPU implementation of MS-REDUCE. The unified memory technique enables quick and easy development of GPU-based algorithms. For our purpose, we simply took the sequential version of G-MSR [4] and modified the code following rules of GPU algorithm development using CUDA unified memory [6]. For scalability study, we appended the UPS2 dataset multiple times to get progressively larger datasets. Figures 8.9 and 8.11 shows that GPU-DAEMON based implementation consistently out-performs the naive implementation. It can be observed in Fig. 8.11 that CUDA unified-memory-based implementation reaches its in-core memory limit at only 14,000 spectra, while G-MSR as shown in Fig. 8.6 reaches its in-core memory limit at 400,000 spectra. Along with better speed, GPU-DAEMON helps conserve limited in-core memory so that more throughput can be achieved. Figure 8.10 shows that GPU-DAEMON version uses a very small amount of in-core memory in comparison to the unified-memory-based implementation.

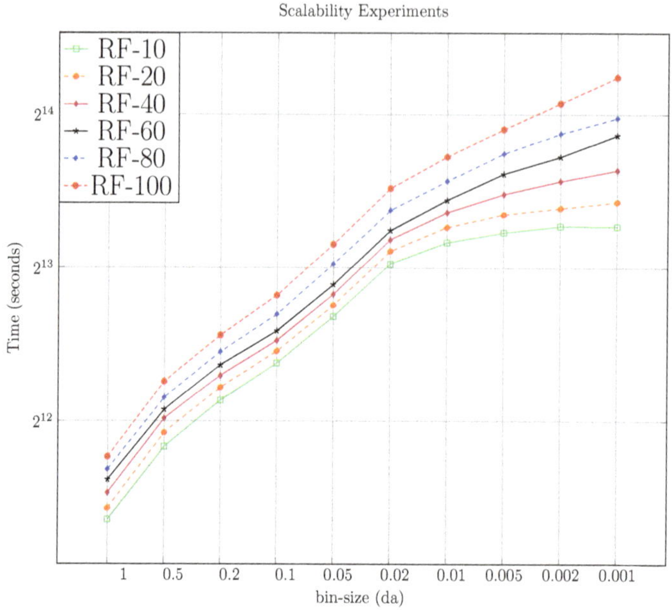

Fig. 8.8 Timing plots of peptide deduction process using Tide with hiXcorr algorithm. Here RF is the reduction factor. An increasing RF makes the process more scalable

Fig. 8.9 Total speed up achieved by GPU-DAEMON implementation over CUDA unified memory-based implementation

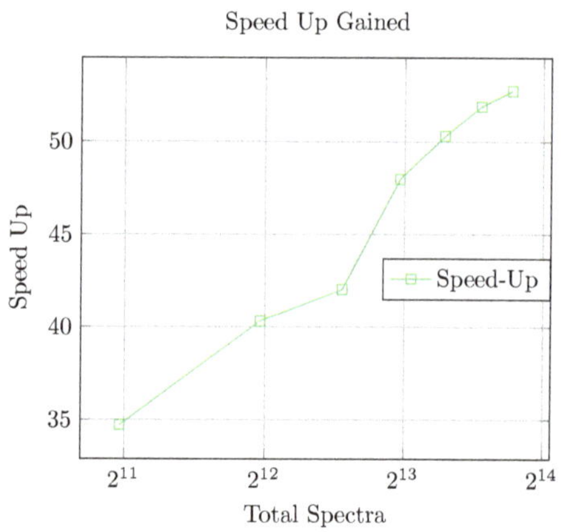

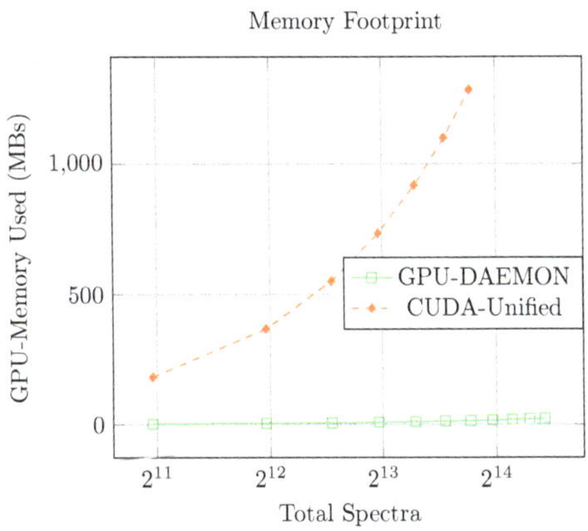

Fig. 8.10 Figure shows that GPU DAEMON based implementation of MS-REDUCE uses only a fraction of memory as used by the CUDA unified memory implementation

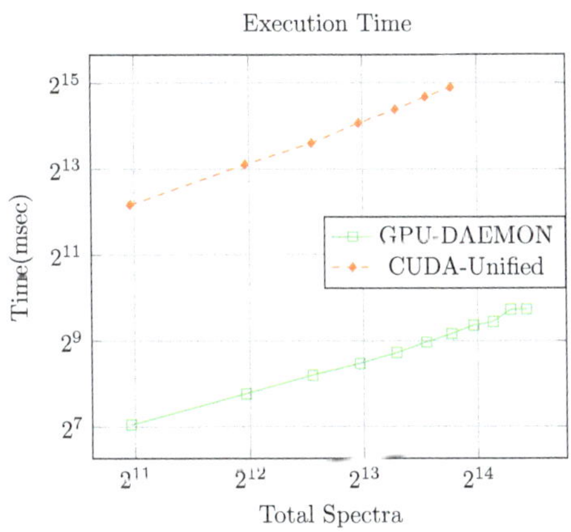

Fig. 8.11 Figure shows that GPU-DAEMON-based implementation of MS-REDUCE scales better with increasing spectra. It should be noticed that CUDA unified memory-based version reaches in-core limit earlier and cannot process more than 14,000 spectra in a single pass while GPU-DAEMON implementation can process about 400,000 spectra before that limit is reached Fig. 8.6

References

1. Mujezinovic N, Schneider G, Wildpaner M, Mechtler K, Eisenhaber F (2010) Reducing the haystack to find the needle: improved protein identification after fast elimination of non-interpretable peptide ms/ms spectra and noise reduction. BMC Genomics 11(1):S13
2. Kong AT, Leprevost FV, Avtonomov DM, Mellacheruvu D, Nesvizhskii AI (2017) Msfragger: ultrafast and comprehensive peptide identification in mass spectrometry-based proteomics. Nature Methods 14(5):513–520
3. Awan MG, Saeed F (2017) An out-of-core GPU based dimensionality reduction algorithm for big mass spectrometry data and its application in bottom-up proteomics. In: Proceedings of the 8th ACM international conference on bioinformatics, computational biology, and health informatics. ACM, pp 550–555
4. Awan MG, Saeed F (2016) Ms-reduce: an ultrafast technique for reduction of big mass spectrometry data for high-throughput processing. Bioinformatics 32(10):1518–1526
5. Baskaran MM, Bordawekar R Optimizing sparse matrix-vector multiplication on GPUS using compile-time and run-time strategies, IBM Research Report, RC24704 (W0812-047)
6. Nvidia (2018) CUDA toolkit documentation. https://docs.nvidia.com/cuda/cuda-c-programming-guide/

Chapter 9
Re-configurable Hardware for Computational Proteomics

Fahad Saeed, Muhammad Haseeb, and Sumesh Kumar

9.1 Introduction

In 1984, when the world was introduced to the first-ever reconfigurable hardware (FPGAs) device, it offered to solve a critical problem faced during the implementation of application specific integrated chips (ASIC). Even though FPGAs ran at a clock speed much slower than that of an ASIC, they provided an attractive solution to emulate the design logic and verify the functional and timing performance at the early stages of the design process. Over the years, FPGAs have evolved from being a tiny fraction of the integrated-circuit manufacturing industry to becoming an industry of their own. With the introduction of advanced lithographic process technologies, it is now cost-efficient to produce multibillion gate FPGAs and deploy larger designs directly in the solutions used in high-performance computing (HPC) industry. This chapter describes the potential of the groundbreaking role that FPGAs can play in developing HPC solutions for the application area of proteomics. The architecture of FPGAs and a number of design strategies using FPGAs for proteomics are discussed in the sections that follow.

9.1.1 Construction of a Field-Programmable Gate Array

A high-level snapshot of a typical FPGA architecture is shown in Fig. 9.1. A mesh of configurable logic blocks (CLB) is connected by programmable routing logic blocks. Each CLB consists of several look up tables (LUT) implementing general purpose logic functions required by any design. The FPGA programming bitstream enables the specific logic paths to achieve a required functionality. The CLBs also contain registers at its output to facilitate the development of pipelined designs.

F. Saeed and M. Haseeb, *High-Performance Algorithms for Mass Spectrometry-Based Omics*, Computational Biology 34,
https://doi.org/10.1007/978-3-031-01960-9_9

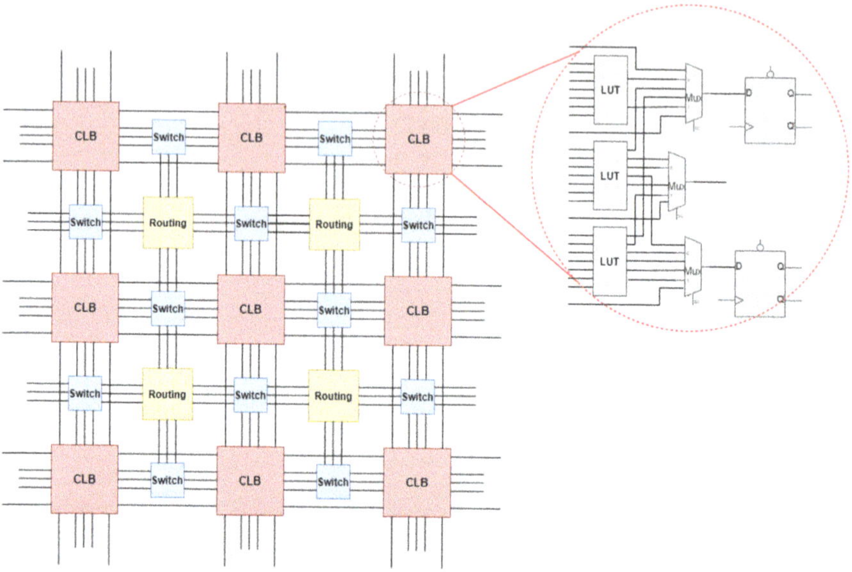

Fig. 9.1 FPGA structure built on an island style routing structure

The FPGA contains an ample amount of routing resource blocks that can be programmed to enable the tristate buffers which connect the wires inside the block. These connections not only enable connections between CLBs but between short segments to form long wire segments as well.

9.2 Popular Architectural Configurations Using FPGAs

After reviewing the intricacies of an FPGA, it is not hard to imagine the kind of applications that will be extremely successful at exploiting this architecture. While ASICs are designed to run at clock speeds reaching in gigahertz, an FPGA can only execute its designs at a few 100 MHz of clock speed . However, the benefit of FPGA based designs comes from its ease of programmability when compared to ASICs, better throughput by developing highly pipelined designs compared with CPUs, and better achievable power efficiency when compared to GPUs. These conclusions have been drawn from works that have attempted to use FPGAs for DSP applications such as matrix multiplication and video processing, and bioinformatics such as string-matching algorithms in genomics. Although different requirements warrant different kinds of design approaches to take, a few of the popular choices of design approach are presented here (Figs. 9.2, 9.3 and 9.4).

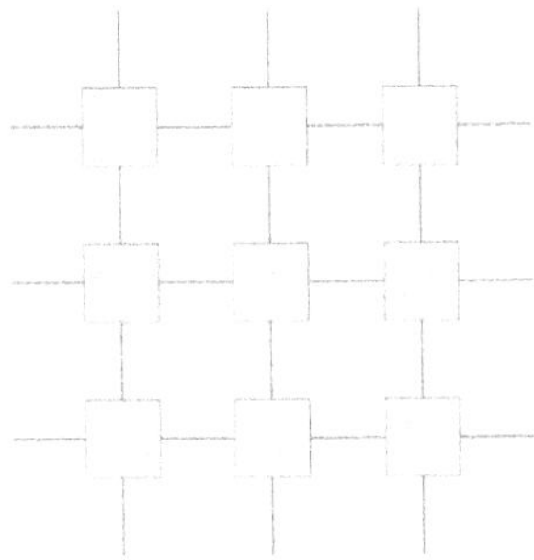

Fig. 9.2 Rectangular mesh systolic array architecture

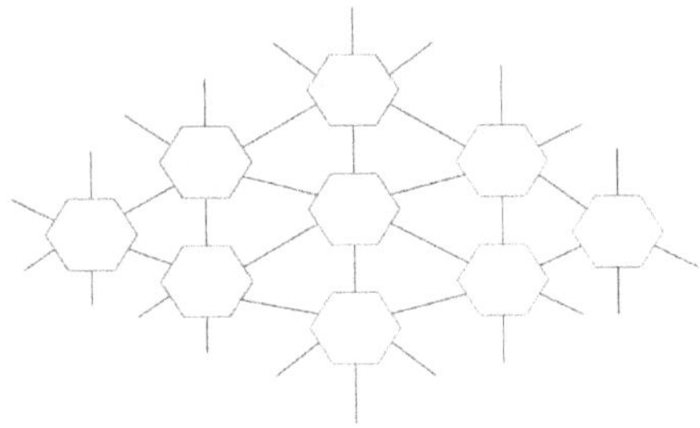

Fig. 9.3 Hexagonal arrangement of processors

Fig. 9.4 Linear systolic array

9.2.1 Systolic Array Configuration

A systolic array configuration refers to the arrangement of large number of relatively simple processors or processing elements (PE) with local interconnects between them. These types of architectures are especially suitable for algorithms with regular dataflow requiring sequential access to the main memory. Various approaches have been explored to define the interconnections between the PEs, a few of which are presented in Figs. 9.2, 9.3 and 9.4.

In a systolic configuration, inputs and outputs can either flow in any direction or they can remain stationary inside the PE. The PEs on the edge is either connected to the memory bus or the input/output collector modules. The entire circuit runs at moderately low clock frequencies compared to an ASIC, but a more efficient use of the FPGA resources is achieved as most PEs are processing on a piece of data simultaneously to increase the system throughput. Different configurations have been experimented for different application problems such as matrix multiplication and convolution. Figure 9.3 shows a hexagonal arrangement where each PE is connected to six neighboring PEs, while each PE is connected to four neighboring PEs in rectangular arrangement, and to two neighboring PEs in a linear arrangement. The trade-off between these choices comes in the form of throughput verses power-consumption. For example, while total number of simultaneously executing PEs might be greater in a hexagonal arrangement, its total number of idle PEs is also much greater compared to a linear arrangement. Another advantage of systolic architecture is the ability to utilize the routing resources on chip in a more efficient manner. If we observe Figs. 9.2, 9.3 and 9.4, the inherent architecture of the FPGA is systolic in nature where each PE logic can be implemented by using one or more adjacent CLBs on the FPGA, and routing resources are used to connect to only neighboring PEs which avoids the use long wire segments and helps to meet timing requirements.

Systolic arrays provide optimal performance if certain conditions are met.

1. The computation time for each PE must be equal as all PEs depend on their neighbors for partial results.
2. Memory access pattern must be non-random or deterministic.

9.2.2 *Parallel Asynchronous PEs Connected to the System Bus*

Since systolic arrays are synchronous in nature, i.e. the computation of PEs must be synchronized, a different approach can be adopted for algorithms having random memory access patterns and irregular computation pattern. Figure 9.5 shows PEs connected to the main system bus and have no interconnections with the other PEs. Each PE starts processing on the data as soon as it receives it. A separate logic block can be designed which only serves to offload the input data to PEs and collect the corresponding outputs. All the PEs in the system execute asynchronously in the sense that they do not need to synchronize their computation with other PEs in the system. However, the performance of such a system is greatly limited as more PEs are used, which cause congestion on the single bus shared by all the PEs. The right number of PEs is selected by calculating the trade-off between memory operation time and computation time. Another challenge is to balance the computation load among the PEs evenly. For computational loads with less degree of data inter-dependency, load balancing can be easily achieved as the workload can be divided by simply distributing (n/p)th of the total input data n among p processing elements.

9.2.3 *Parallel Processors with Communication Interconnect*

Another architectural setting which combines the features of systolic and asynchronous processing features is shown in Fig. 9.5. This kind of configuration allows local processing which is independent of the execution of other PEs in the system. However, it includes communication interconnect between the processing elements to allow the sharing of computation parameters and input data processing (Fig. 9.6).

The PEs are connected to the main memory bus which is accessible by the host CPU. For the host CPU, PEs act as co-processor units for carrying out computation

Fig. 9.5 PE is represented by gray block. Each PE executes independently of other PEs in the system

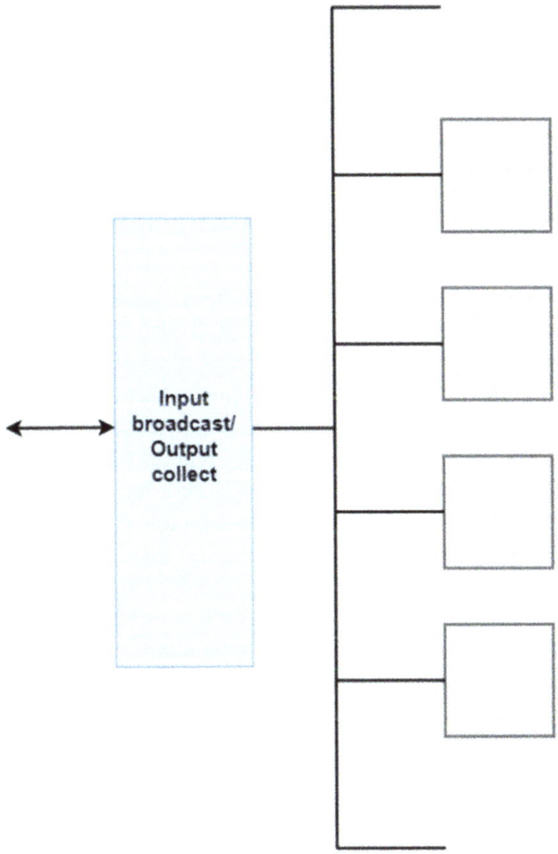

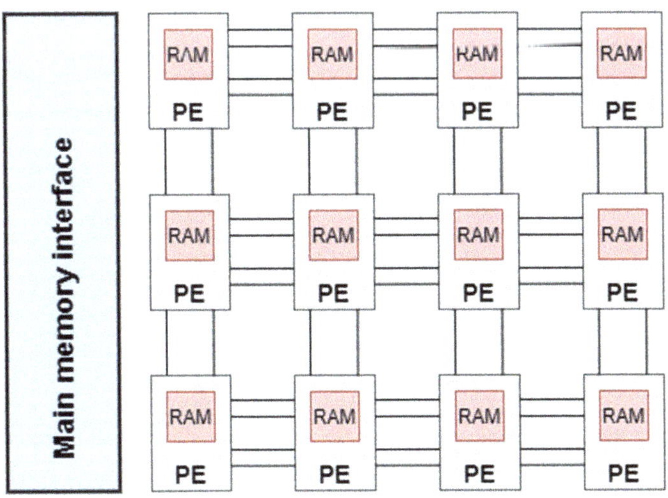

Fig. 9.6 Parallel processors with intercommunication network to share features and data

and CPU is responsible for orchestrating the flow of computation as the nature of computations mostly being performed are irregular in nature. Each PEs has its own local RAM which allows execution of light weight computation kernels and local data management rather than simple compute operations as we saw in the pure systolic array architectures.

9.3 FPGA Design for Computational Proteomics

The most demanding computational problems in proteomics are those involving peptide identification. In the peptide identification process, millions of experimental spectrum vectors are compared against a peptide database of already discovered peptides. Some of the most popular database search algorithms rely on computing a similarity score between an experimental spectrum vector acquired from mass spectrometers and a theoretically generated spectrum from the peptide. The similarity scores basically provide a quantitative measure of the number of peaks that match between the experimental and theoretical spectrum. One such similarity score is called Xcorr which is computed by taking the cross-correlation between the two vectors. Following the main principles of high-performance architecture design using FPGAs described above, an example architecture for accelerating the computation of Xcorr is presented here. The similarity score between a theoretically generated spectrum vector X and an experimentally acquired spectrum vector Y of length n is defined as

$$X_{corr} = \sum_{i=0}^{n-1} X[i]Y[i] - \frac{1}{151} \sum_{i=0}^{n-1} \sum_{\tau=-75}^{\tau=75} X[i]Y[i-\tau] \tag{9.1}$$

Equation (9.1) calculates the difference between the dot product and cross-correlation calculated with an offset value of $\tau = 75$. This calculation can be simplified by performing a pre-processing step which allows us to carry out the expensive dot product operation only once instead of performing dot products across all offsets. The pre-processing step is summarized below,

$$X_{corr} = \sum_{i=0}^{n-1} X[i] \left(Y[i] - \frac{1}{151} \sum_{\tau=-75}^{\tau=75} Y[i-\tau] \right)$$

$$Y_P = \sum_{i=0}^{n-1} \left(Y[i] - \frac{1}{151} \sum_{\tau=-75}^{\tau=75} Y[i-\tau] \right) \tag{9.2}$$

by following the simplification in (9.2), we can reduce the overall computation to

$$X_{corr} = \sum_{i=0}^{n-1} X[i] Y_P[i]$$

9.3.1 Architecture Overview

A bird's eye view of the architecture design to implement Eq. (9.1) is shown in Fig. 9.7. In this setup, the host CPU acts as a computation manager, whose only job is to transfer the required mass spectrometry data to the FPGA DRAM via the PCIe link. On the FPGA side, the PCIe communication between CPU and the logic is performed via the PCIe DMA. Along with that, FPGA logic has a lightweight core defining a register file to hold the parameters of computation and communicate with the CPU. Other important components in the design are the processing element (PE) and the bus-arbitration module.

The main steps of the algorithm—reading experimental spectra from the DRAM , candidate peptide search, theoretical spectra ion generation, dot product computation, and compiling resulting scores—are carried out by the PE. In this design setting, there can be as many PEs executing in parallel as allowed by the number of resources available on the FPGA fabric. All the PEs execute the computation independently of each other and on their local copy of experimental spectrum information. Since there is only one DRAM, all the PEs have access to it via the system bus. A bus-arbitration module sits between the system bus and the PEs which allocates access to the PEs based on a fairness scheme. In this scenario, where the size of request (size of data to be transferred) will be roughly equal for all the PEs, one fair scheme which is implemented here is the first-come-first-serve (FCFS) policy. The following sub-sections explain the construction of some of the important sub-modules in the design.

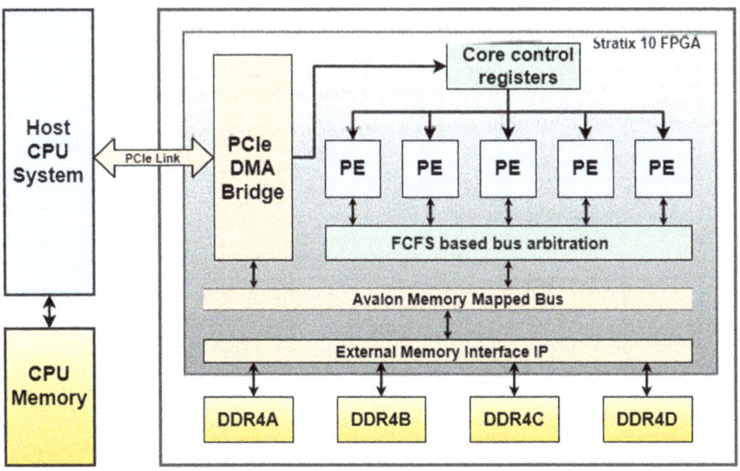

Fig. 9.7 Complete system architecture shows host CPU communicates with FPGA RAM via PCIe DMA bridge which is connected to Intel's Avalon memory mapped bus. Core registers module contains the computation parameters and is also used for FPGA-CPU communication. To allow efficient use of the Avalon memory mapped bus, all PEs are connected to FCFS-based bus arbiter which is in turn connected to Avalon memory mapped bus

9.3.2 Processing Element (PE)

The processing element is in essence a lightweight single-purpose processor whose instructions are the computation parameters which are stored in the core-compute registers. The flow of computation in this algorithm is maintained by a state machine which carries out all the steps in order. In Fig. 9.8, the state machine is represented by the block called "Control Logic". The control logic performs the following steps in order

1. Request access to the bus arbiter for memory access.
2. After access is granted, copy the next available spectrum from the DRAM into the local on-chip RAM.
3. Read the precursor mass of the copied spectrum and activate binary search module to find candidate peptides.
4. Copy the candidate peptides into the peptide registers.

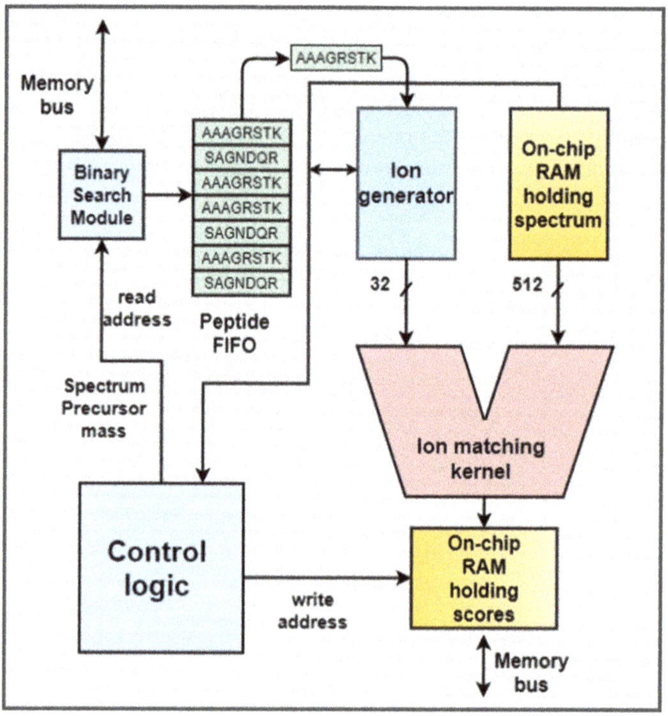

Fig. 9.8 Detailed internal construction of a single processing element. At the heart is the control logic which controls the function of all the sub-modules in the figure. Binary search module fetches a candidate peptide and stores it in a peptide FIFO. Ion generator reads the peptide and generates fragment-ions. A 512-bit packet containing 16 32-bit ion-mz and intensity pair values along with a 32-bit theoretical ion and intensity pair values are fed to the ion-matching kernel which finds the matching peak and stores the partial score in on-chip RAM

5. Generate the predicted peptide ions theoretically.
6. Read the experimental spectrum ions and theoretically generated ions and pass them to ion-matching module.
7. Wait for the ion-matching module to finish the computation of dot product score.
8. After completion, the resulting scores for the current experimental spectrum in local RAM are placed into the result RAM.
9. Read the dot product scores along with experimental spectrum information and transfer them to the DRAM.
10. Signal the address of the resulting scores and the completion of operation to the CPU by writing to core control registers.

9.3.3 *Bus-Arbitration Module*

The bus-arbitration module ensures that all the PEs in the system have an equal amount of workload so that maximum performance benefit is achieved. In the case when a large number of PEs are connected to the system bus, a huge number of requests can congest the bus resulting in uneven bus allocation. The bus arbiter module avoids this by keeping track of the order in which access requests were received. In this way, PEs are assigned based on a first-come-first-serve (FCFS) policy. The detailed architectural view of the bus arbiter module is shown in Fig. 9.9. The arbiter module has input lines named "bus request" which are coming from the PEs. Whenever a PE generates the request for access to the bus, the wait register connected to that signal counts and keeps track of the time elapsed since the request was generated. When the bus finally becomes available, a find-max module finds the register which has the largest wait time value, and the arbiter assigns the access of the bus to the corresponding PE.

9.3.4 *Binary Search Module*

The peptides are placed in DRAM sorted by their mass value. This facilitates the searching for candidate peptides as the search key is also based on the precursor mass value for the experimental spectrum. The search operation is performed in $\log(n)$ time—where n is the total number of peptides—by the binary search module. The hardware circuit to perform a binary search operation is shown in Fig. 9.10. For every request, the search module starts by accessing the memory location at $n/2$ and compares the mass of the read peptide with the mass of experimental spectrum. In the circuit, the address register is used to hold the address to the current location being accessed. If the mass lies in the precursor window, the location of the peptide is returned, otherwise the address register is updated to search in the location $n/2 + \delta$ or $n/2 - \delta$, based on the previous comparison. This operation is repeated until the required precursor window condition is satisfied.

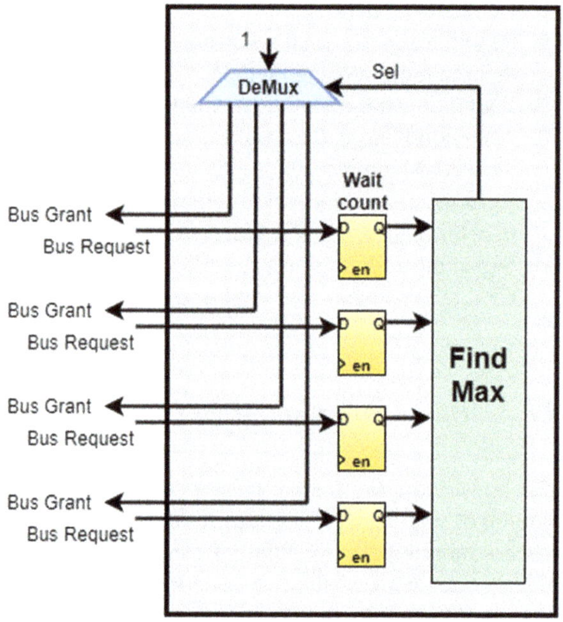

Fig. 9.9 Brief logic description of bus-arbitration module. Bus request lines from all the PEs are coming into the arbiter. When a PE is denied service, its wait count register is incremented, dynamically increasing its priority for the next turn. Find max module is a comparator tree that finds which registers have the maximum value and grants access to the corresponding PE

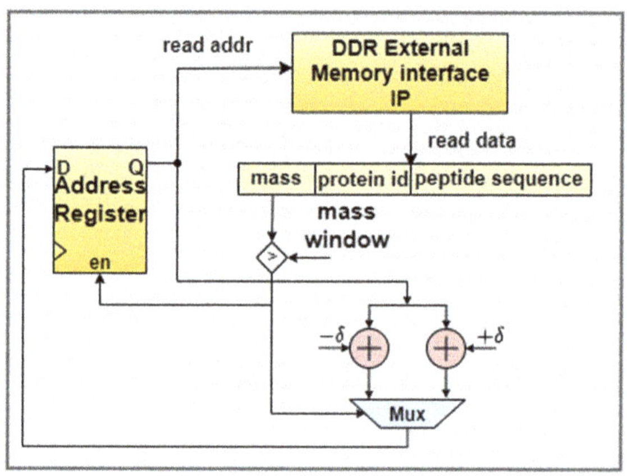

Fig. 9.10 Control logic to perform binary search for the precursor-ion mass. Based on the precursor mass window, search module start from address location $n/2$ where n is the last address. A MUX selects the next address to be read if a candidate peptide is not found

9.3.5 Ion-Matching Circuit

Prior to the beginning of computation, the experimental spectrum vector is stored in the local on-chip RAM in the compressed sparse row (CSR) format. The ion-matching module is a combinational circuit which is connected to the experimental spectrum vector in the on-chip RAM and the theoretically generated ions coming from the ion-generator module. The computation begins by moving packets of experimental spectrum pairs of ion m/z and intensity values into the circuit. Each packet is a

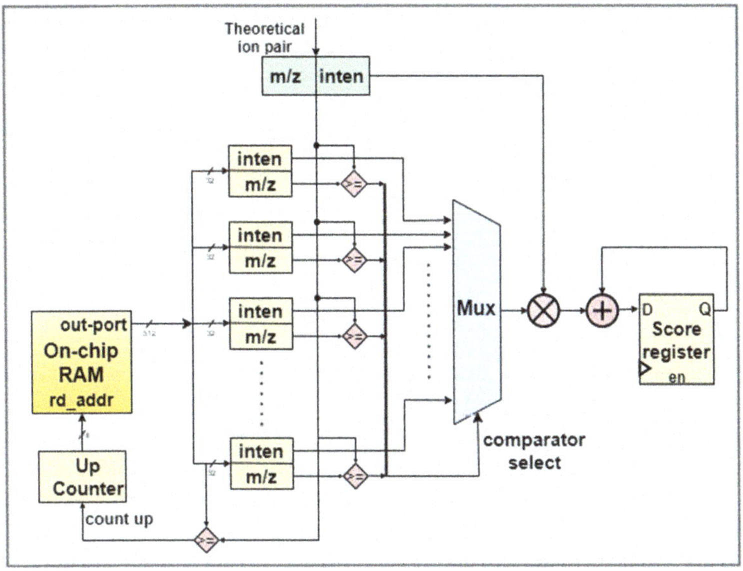

Fig. 9.11 The ion-matching circuit receives a 512-bit packet containing 16 experimental ions which are all compared with a theoretical ion in one cycle. The matched ions are multiplied and accumulated in the score register. If the theoretical ion is outside the range of current experimental ions, next packet is requested from the on-chip-RAM by incrementing the counter

64-byte long word containing 16-ion pairs. The packet size was determined from the DRAM read packet size which is 6-byte long. The intensity value of the 16 ions is compared with a theoretical ion to check if it lies in range. This comparison is done using 16 parallel comparators as shown in Fig. 9.11. If the theoretical ion is matched, the corresponding intensity value is multiplied by the intensity value of the theoretical ion and accumulated in the score register. If the ion is not matched, then either a new theoretical ion is generated by the module or a new experimental spectrum vector is read from the on-chip RAM. In this manner, the ions from both experimental and theoretical vector are matched sequentially to finish the dot product scoring. After completion of the computation, the final score is stored in a separate on-chip RAM collecting results.

9.3.6 Experiments and Results

The hardware design explained above was designed using Intel Qsys system builder and Quartus for Stratix 10 FPGA board. The design was tested for a clock frequency of 200 MHz. The dataset used for this experiment was from PRIDE (Proteomics Identification Database) called PXD000612 containing more than 90,000 experimentally acquired spectra which were matched against more than 669k peptides from the

Table 9.1 Total processing time for 90494 experimental spectra compared with 669964 peptides. Here the time is shown for the number of PEs instantiated in the system verses cache size used

Cache	Total processing time for #PEs (s)							
Size	1	4	8	12	16	20	24	32
512B	54.63	16.65	15.43	15.25	15.28	15.10	15.03	14.98
1kB	48.01	12.36	9.36	8.57	8.18	8.06	7.84	7.76
2kB	39.44	7.87	4.35	3.02	2.30	1.84	1.55	1.25
4kB	37.62	7.42	4.11	2.84	2.17	1.75	1.47	1.18

human proteome dataset. For this search process, there were over 316 million dot product computations performed.

The design achieves its performance gains by optimizing the number of DRAM accesses and employing input reuse. In digital systems, main memory accesses are much more expensive compared to data access from on-chip resources. In this design, input reuse is implemented by carefully selecting the size of on-chip RAM which acts as processor cache. In computer systems, caching works by predicting the data which will be accessed next so that it is already available next to the processor instead of waiting for it to arrive from the DRAM. In this design, there is no need of predicting as it is already known that is expensive to copy from the DRAM in experimental spectrum vectors. Each experimental vector has to be comparted to all the candidate peptides separately, thus in order to minimize DRAM accesses it is imperative to copy it only once and keep it in cache until it is no longer be required for future computations. If the cache size is not big enough to hold the entire spectrum, then it must be copied from the DRAM in chunks. Since each chunk must be discarded after it has been processed, it has to be copied again for every candidate peptide. This will result in exponentially higher number of DRAM accesses compared to the case when cache size is big enough to hold the entire spectrum vector. The correct size of cache was chosen by experimentation on the dataset. The results of this design space exploration are illustrated in Table 9.1. Table 9.1 shows the effect of cache (implemented as on-chip RAM) size on the above-defined three parameters. It is evident that the total computation time decreases linearly with increasing the number of processing elements when the cache size is 2 kB or 4 kB. However, for a cache size below 2 kB, the speedup does not follow a linear behavior after increasing the PEs above 6.

The analysis was extended to clearly demonstrate the time spent on computation and communication, where the communication time is further divided in terms of time spent on I/O (memory operations in this case) and waiting time. These terms are defined as

Average compute time: Timc spent by the processor on computing dot product.

Average I/O time: Time spent by the processor on performing memory read/write operations.

Average wait time: Time spent by processor on waiting to get access to the memory bus.

Table 9.2 Average wait time of each processing element for number of PEs instantiated in the system verses the size of cache used

Cache	Average wait time for #PEs (s)							
	1	4	8	12	16	20	24	32
512B	0.0	5.78	9.38	15.25	11.08	11.93	12.47	13.13
1kB	0.0	2.80	4.07	4.91	5.40	5.81	5.97	6.26
2kB	0.0	0.0018	0.0020	0.0022	0.0022	0.0023	0.0023	0.0023
4kB	0.0	0.0018	0.0019	0.0020	0.0021	0.0021	0.0021	0.0022

Table 9.3 Average time spent by each processing element on performing DRAM read/write operations for number of PEs instantiated in the system verses size of cache used

Cache	Average I/O time for #PEs (s)							
	1	4	8	12	16	20	24	32
512B	15.39	3.05	1.69	1.17	0.88	0.71	0.59	0.48
1kB	8.11	1.60	0.88	0.60	0.46	0.37	0.30	0.24
2kB	0.0081	0.0016	0.0009	0.0007	0.0006	0.0004	0.0003	0.0002
4kB	0.0077	0.0015	0.0008	0.0005	0.0004	0.0003	0.0002	0.0002

Tables 9.2, 9.3, and 9.4 show the breakdown of total time spent in terms of the above-defined terms. These numbers represent the average time spent by a single processing element in the system. In Table 9.3, we can see the pattern of change in waiting time when a cache size of 512 B, 1 kB, 2 kB, and 4 kB is used in the system. When the cache size is below 2 kB, the average wait time is shown to be increasing exponentially when the number of PEs is increased. However, when a 2 kB or 4 kB of cache is used, the wait time stays roughly constant for up to 32 PEs.

Table 9.3 shows a similar trend in average I/O time, there is an exponential increase in the time spent on read/write operations with the DRAM when the cache size is below 2 kB. The time spent on memory operations when cache is below 2 kB is 600x higher. This was expected as we know that the size of largest experimental spectrum vector in our dataset is less than 2 kB. So, when the cache is not large enough to keep the entire spectrum vector, then the PE has to copy the vector in chunks, and it has to copy the same vector all over again for every candidate peptide.

Table 9.4 demonstrates an important result, i.e. the average amount of time spent by a processing element on performing dot product is not affected by the size of cache. For any cache size, the computation time scales linearly with the number of processing elements instantiated in the system irrespective of the size of cache used. This is in line with the design decision described in earlier sections, i.e. the total processing time in this problem is stalled due to the inefficient memory access operation.

Table 9.4 Average time spent by each processing element on performing dot product computation operations for number of PEs instantiated in the system verses size of cache used

Cache	Average compute time for #PEs (s)							
	1	4	8	12	16	20	24	32
512B	39.23	7.81	4.35	3.00	2.28	1.85	1.54	1.24
1kB	39.89	7.95	4.40	3.05	2.31	1.87	1.56	1.26
2kB	39.43	7.86	4.34	3.01	2.29	1.84	1.55	1.25
4kB	37.25	7.42	4.11	2.84	2.16	1.75	1.47	1.17

9.4 Conclusion

The design presented in this chapter to accelerate the computation of Xcorr using FPGAs provides some insight on performing architecture level optimizations to create an efficient communication and computation pipelines. Using the high reconfigurability of FPGAs, a design can be adapted to any kind of requirements demanded by the application. Moreover, performing design space exploration by measuring performance metrics against varying important system parameters such as the number of processing elements and cache size, etc., one can find the most optimal choice of parameters in the final design.

Chapter 10
Machine-Learning and the Future of HPC for MS-Based Omics

Fahad Saeed and Muhammad Haseeb

To date, MS proteomics data is identified using database search algorithms based purely on numerical techniques or some denovo techniques that allow peptide identification without using databases. Currently, there is no single strategy from database search or denovo techniques that can claim as the most accurate strategy. Substantial work has been carried out toward developing computational techniques for identification of peptides using database search [1], as well as denovo algorithms [2]. However, peptide identification problems are well-known and prevalent [3] including but not limited to misidentifications or no identification for peptides, statistical accuracy (FDR) and inconsistencies between different search engines [4].

Most of the algorithms applied to MS data have been limited to traditional numerical algorithms and can be categorized as database search algorithms and denovo algorithms. Comparison across literature indicates decreased average accuracy of denovo algorithms (38.1–64.0%) [4] relative to database search algorithms (30–80%) [5]. However, within-study direct comparisons of database verses denovo machine learning (ML) approaches have revealed modest gains [6], indicating further formal evaluation is warranted. Moreover, ML methods use different validation metrics and the lack of standard metrics and/or data-benchmarks can lead to overly optimistic assessment for machine learning algorithms [7]. Overall, prior literature demonstrated limited accuracy and generalizability [4] identifying peptides using current limited ML methods.

Previous work suggests that numerical algorithms and the use of traditional ML algorithms may not be able to capture and integrate the multidimensional features of MS data [8]. However, deep learning methods [4, 8] may offer an improved approach for identifying peptides in noisy high-dimensional MS data and peptides that are very similar to each other [9]. Preliminary progress assessing deep learning methods in peptide deduction applied to MS data has yielded an average accuracy of 82–95% on selected data sets but with limited precision (amino acid level 72%)

F. Saeed and M. Haseeb, *High-Performance Algorithms for Mass Spectrometry-Based Omics*, Computational Biology 34,
https://doi.org/10.1007/978-3-031-01960-9_10

and recall (peptide—level 39.24%) [8]. However, large volumes of data with large number of possible parameters are needed for deep learning training, particularly for MS data, which has resulted in a technical hurdle in developing such strategies. We have previously shown that this can result in overfitted deep learning models [10] with limited increase in accuracy due to noise feature-integration. Such overfitting leads to limited generalizability [11], and contributes to the ongoing reproducibility crisis [12–14]. One Deep-learning algorithm when used on another's data set leads to like 30% accuracy which suggests that there is a generalizability problem [15, 16]. Further, existing ML algorithms are computationally expensive and subsequently have limited scalability in training and application. Our prior work has demonstrated that this is particularly true for MS applications that scale poorly with increasing size of data sets [17]. Computational scaling and management are needed for these machine learning algorithms as this is currently a significant challenge for proteomics practitioners interested in applying these techniques.

In the future, we foresee the integrated use of image-processing, machine learning, including deep learning for MS data, to identify peptides from MS data in a highly accurate manner. To this end, there is some literature that has focused on processing MS data using machine learning and deep learning techniques. Importantly, image-processing, deep learning strategies, and fusing of multi-modal features have not been applied to MS data even though these techniques have the potential to radically change how MS data is processed with highly accurate peptide identifications. However, to make the proposed deep learning training and solutions scalable, HPC algorithms are needed.

10.1 Why HPC is Essential for Machine-Learning Models

Deep-learning models have unmatched expressive power as compared to traditional machine learning solutions. However, this expressive power comes from very large number of trainable parameters which can capture complex relationships between the data. In general, bigger and deeper Convolutional Neural Networks (CNN) models are used for various applications that are successful. The objective of this aim is to design and develop high-performance computing strategies which can process big MS data and accelerate the proposed deep learning models. CPU-GPU-based methods can give superior speeds in the processing of MS data, and of training, and inferring of deep learning models. Successful completion of such CPU-GPU-based pipelines is likely to contribute fundamental HPC techniques to our base of knowledge, without which the complex training and inferring of deep learning networks cannot be accomplished in reasonable timeframes. Upon completion of this research challenge, it is our expectation that we will have developed a HPC framework for the proposed DL solutions for MS-based omics computational strategies developed by us/others. Such tools would be important because they would likely aid in much-needed approaches to study human gut/environments microbiomes in a scalable fashion.

As part of our preliminary studies (Haseeb et al. Nature 2021), we explored our recently proposed HPC framework (called HiCOPS) [18] for efficient acceleration of database peptide search algorithms on large-scale symmetric multiprocessor distributed-memory supercomputers. HiCOPS exhibits orders-of-magnitude improvement in speed compared with several existing shared- and distributed-memory database peptide search tools, allowing several gigabytes of experimental MS/MS data to be searched against terabytes of theoretical databases in a few minutes compared with the several hours required by existing algorithms. The proposed HiCOPS parallel design implements an unconventional approach in which the (massive) theoretical databases are distributed across parallel nodes in a load-balanced fashion followed by asynchronous parallel execution of the database peptide search. On completion, the locally computed results are merged into global results in a communication-optimal manner. We have demonstrated an extensive performance evaluation in which we report between 70 and 80% strong-scale efficiency and less than 25% overall performance overheads (load imbalance, I/O, interprocess communication, pipeline halt); collectively depicting a near-optimal parallel performance. This overhead cost-optimal design, along with several optimizations, allows HiCOPS to maximize resource utilization and alleviate performance bottlenecks. Since our HPC framework is search-algorithm oblivious it will be a natural extension to incorporate meta-proteomics deep learning model into the HPC framework.

10.2 Preliminary Data and Findings

In our recent paper [19], we have designed and implemented a Deep Cross-Modal Similarity Network called SpeCollate. This is a deep learning network that tries to learn the scoring function between the spectra and peptides by mapping the different modalities of the data into a shared Euclidean subspace. This is achieved by learning fixed sized embeddings, and training the network using sextuplets of positive and negative examples. SpeCollate also uses a custom-designed SNAP-loss function and hardest negative mining for appropriate negative examples to improve the training performance. In order to train the network 4.8 million sextuplets obtained from the NIST and MassIVE peptide libraries were used and which allowed our deep learning model to perform better than Crux and MSFragger in terms of the number of peptide-spectrum matches (PSMs) and unique peptides identified under 1% FDR for real-world data. To the best of our knowledge, our deep learning network is the first model that can determine the cross-modal similarity between peptides and mass spectra for MS-based omics.

Despite all the superior accuracy of the deep learning model one thing that was a severe bottleneck was the time it takes to train the network. Since there is no theoretical framework that would allow us to predict the performance of the model, one has to *fully* train the model and run validation and testing before any judgement can be made. Our single preliminary design shows that it takes approx. 283 days of compute time to train a single deep learning network with 5 hyper-parameters optimizations. The intractability of a single model demonstrates that continued search for better DL models is huge technical hurdle in studying MS-based meta-omics. Each possible deep-network, even though carefully selected, has to be completely trained to assess its feasibility. Apart from the time it takes for the deep learning model training and inferences; the workflows that process MS data are also shown to be inefficient, both theoretically [20] and in real-world experiments [18, 21]. We believe, the HPC frameworks, that can accelerate the MS data analysis as well as the training and inference of deep learning will be essential to tract any kind of MS-based omics data analysis.

We believe that HPC will be at the forefront of these scientific investigations in the future.

References

1. Deutsch EW, Mendoza L, Shteynberg D, Slagel J, Sun Z, Moritz RL (2015) Trans-proteomic pipeline, a standardized data processing pipeline for large-scale reproducible proteomics informatics. PROTEOMICS-Clin Appl 9(7–8):745–754
2. Vyatkina K, Wu S, Dekker LJ, VanDuijn MM, Liu X, Tolic N, Dvorkin M, Alexandrova S, Luider TM, Pasa-Tolic L et al (2015) De novo sequencing of peptides from top-down tandem mass spectra. J Proteome Res 14(11):4450–4462
3. Griss J, Perez-Riverol Y, Lewis S, Tabb DL, Dianes JA, del Toro N, Rurik M, Walzer M, Kohlbacher O, Hermjakob H et al (2016) Recognizing millions of consistently unidentified spectra across hundreds of shotgun proteomics datasets. Nat Methods 13(8):651
4. Tran NH, Zhang X, Xin L, Shan B, Li M (2017) De novo peptide sequencing by deep learning. Proc Natl Acad Sci 114(31):8247–8252
5. Chick JM, Kolippakkam D, Nusinow DP, Zhai B, Rad R, Huttlin EL, Gygi SP (2015) A mass-tolerant database search identifies a large proportion of unassigned spectra in shotgun proteomics as modified peptides. Nat Biotechnol 33(7):743
6. Chi H, Liu C, Yang, H Zeng W-F, Wu L, Zhou W-J, Niu X-N, Y-H Ding, Zhang Y, Wang R-M et al (2018) Open-pfind enables precise, comprehensive and rapid peptide identification in shotgun proteomics, bioRxiv 285395
7. Obermeyer Z, Emanuel EJ (2016) Predicting the future-big data, machine learning, and clinical medicine. N Engl J Med 375(13):1216
8. Qiao R, Tran NH, Li M, Xin L, Shan B, Ghodsi A Deepnovov2: better de novo peptide sequencing with deep learning. arXiv:1904.08514
9. Zhou X-X, Zeng W-F, Chi H, Luo C, Liu C, Zhan J, He S-M, Zhang Z (2017) pdeep: predicting ms/ms spectra of peptides with deep learning. Anal Chem 89(23):12690–12697
10. Eslami T, Saeed F (2018) Similarity based classification of adhd using singular value decomposition. In: Proceedings of ACM international conference on computing frontiers (CF'18), pp 19–25
11. Zou L, Zheng J, Miao C, Mckeown MJ, Wang ZJ (2017) 3d CNN based automatic diagnosis of attention deficit hyperactivity disorder using functional and structural MRI. IEEE Access 5:23626–23636
12. Allison DB, Shiffrin RM, Stodden V (2018) Reproducibility of research: issues and proposed remedies. Proc Natl Acad Sci 115(11):2561–2562
13. Hutson M (2018) Artificial intelligence faces reproducibility crisis. Science (New York, NY) 359(6377):725
14. Berrar D, Dubitzky W (2017) On the Jeffreys-Lindley paradox and the looming reproducibility crisis in machine learning. In: 2017 IEEE international conference on data science and advanced analytics (DSAA). IEEE pp 334–340

15. Gabriels R, Martens L, Degroeve S (2019) Updated ms^{2}pip web server delivers fast and accurate ms^2 peak intensity prediction for multiple fragmentation methods, instruments and labeling techniques. Nucleic Acids Res 47(W1):W295–W299
16. Gessulat S, Schmidt T, Zolg DP, Samaras P, Schnatbaum K, Zerweck J, Knaute T, Rechenberger J, Delanghe B, Huhmer A et al (2019) Prosit: proteome-wide prediction of peptide tandem mass spectra by deep learning. Nat Methods 16(6):509
17. Haseeb M, Afzali F, Saeed F (2019) Lbe: A computational load balancing algorithm for speeding up parallel peptide search in mass-spectrometry based proteomics. In: IEEE international parallel and distributed processing symposium workshops (IPDPSW). IEEE, pp 191–198
18. Haseeb M, Saeed F (2021). Source data: high performance computing framework for tera-scale database search of mass spectrometry data. https://doi.org/10.5281/zenodo.5076575
19. Tariq MU, Saeed F (2021) Specollate: deep cross-modal similarity network for mass spectrometry data based peptide deductions. PLoS ONE 16(10):e0259349
20. Saeed F, Haseeb M, Iyengar S Communication lower-bounds for distributed-memory computations for mass spectrometry based omics data. arXiv:2009.14123
21. Kumar S, Saeed F (2021) Communication-avoiding micro-architecture to compute xcorr scores for peptide identification. In: 2021 31st international conference on field-programmable logic and applications (FPL). IEEE, pp 99–103

Glossary

Accuracy (peptide identification) The fraction of experimental MS/MS spectra correctly matched to their peptide sequence.

Amdahl's Law A formula to compute the theoretical maximum achievable speedup in parallel computing.

Application Specific Integrated Chip A computer chip or a microprocessor specifically designed to (optimally) perform a certain use in contrast to a general purpose CPU.

Arithmetic operations CPU operations of arithmetic nature including addition, subtraction, multiplication, division, multiply-add etc.

ASIC Application Specific Integrated Chip.

Asynchronous parallel A parallel computing configuration where multiple CPUs or nodes independently and simultaneously work to complete a task or algorithm.

Background noise The noisy MS/MS data introduced by the environment including any impurities, the equipment, or stray ionization in mass spectrometry.

Bandwidth cost The amount of bandwidth (bits/second) consumed by a certain task/algorithm.

Bitstream A sequence of computer bits.

Bolt A parallel computing based algorithm for database peptide search.

Bulk Synchronous Parallel (BSP) Bulk Synchronous Parallel. A parallel computing scheme where a set of asynchronous parallel nodes execute a bulk of algorithmic work before synchronizing in contrast to synchronizing after each cycle or step.

Buckets Set of (similar) items having similar properties or requiring the same computation.

Cache A fast memory commonly used to store frequently used items for the processor.

Chromatogram A color pattern formed by performing liquid chromatography where each color represents a separated compound in the mixture.

CID Collision Induced Dissociation. A method to dissociate peptides or protein ions into fragments.

Co-eluting peptides Peptides that elute together in the mass spectrometer commonly because they are isobaric and are not separated only based on their relative mass.

F. Saeed and M. Haseeb, *High-Performance Algorithms for Mass Spectrometry-Based Omics*, Computational Biology 34,
https://doi.org/10.1007/978-3-031-01960-9

Coalesced memory access A technique to combine several many and haphazard memory accesses into one transaction to improve bandwidth.

Coding genes Genes (regions of a genome sequence) that translate to proteins.

Communication cost Overhead cost of communicating between parallel running CPUs or nodes to synchronize and/or exchange information.

Connectomics Connectomics is the production and study of the connections within an organism's brain and nervous system.

Convolutional Neural Networks (CNNs) CNNs or ConvNets are a class of deep learning that apply sliding kernels to detect features. CNNs have wide application in image and video applications.

Correlation A statistical measure of the degree to which two or more variables are coherent to one another.

CPU-GPU architectures Computer architectures incorporating both general purpose CPUs, Graphical Processing Units (GPUs) and the interconnect between them.

Cross-correlation A method to compute similarity between experimental and theoretical MS/MS spectra.

Curve fitting Fitting a standard curve given by a polynomial to a given chart/ histogram by optimizing polynomial's parameters, coefficients and exponents.

Dark-data Experimental MS/MS spectra that do not match to any peptide sequence and hence are unknown.

Data Dependent Acquisition (DDA) A mode of data collection in mass spectrometry. In DDA, the mass spectrometer selects only the most intense peaks or peptide ions in the first stage, and further fragments them for analysis.

Data Independent Acquisition (DIA) A mode of data collection in mass spectrometry. In DIA, the mass spectrometer divides the peptide precursor mass range into mass windows. Then during each cycle of 2–4 s, the mass spectrometer fragments all peptide ions precursors in the windows.

Database Search A method to identify peptides from the experimental mass spectrometry data. In database search, the experimental mass spectra are compared against a database of theoretical spectra to find the best match. The theoretical spectra are generated using in-silico simulation process using protein sequences.

de novo A method to directly analyze the experimental mass spectrometry data and infer the peptide sequence(s) it may correspond to.

Data parallelism A parallel computing technique where several CPUs perform work by simultaneously performing the same computation on different data.

Data partitioning The process of partitioning a large chunk of data into smaller chunks for simultaneous processing.

Denoising The process of removing spurious data points and noisy peaks introduced by the environment in the actual data.

Dimensionality reduction The process of reducing the number of dimensions (properties) per data point. For example, reducing data points from (x, y, z) → (x, y).

Edit Distance A method to measure the dissimilarity between two sequences or strings.

Embarrassingly parallel An ideal parallel computing situation where increasing the number of nodes or processors by $K\times$ results in a speedup of $K\times$.

Exon A portion of the gene sequence that codes for amino acids.

Intron A portion of the gene sequence that does not code for amino acids.

fMRI Functional magnetic resonance imaging (fMRI) measures brain activity by detecting changes associated with the blood flow.

GPU Graphical Processing Unit.

Heterogeneous computing In heterogeneous computing, the algorithmic workload is parallelized over asymmetric computing resources, often general purpose CPUs and GPUs, in a way to better harness the system to achieve speedup.

High-performance computing (HPC) High-performance computing refers to using supercomputers and large-scale computers to solve computational and numerical problems, often performing massive amounts of complex computations on complex and large data sets, to compute the solution.

Hybrid parallelism A parallelization technique where both the task and data are split among parallel machines to achieve parallelism.

Hyperscore An algorithm to compute similarity between an experimental and a theoretical MS/MS spectrum.

Intragenic Being or occurring within a gene.

Isotopic clusters The set of closely located (clustered) peaks in a spectrum representing different isotopes of the molecules inside the mass spectrometer.

Lipidomics A life sciences sub-field involving large-scale studies of networks and pathways in cellular lipids in biological systems.

LBE A parallel computing algorithm for balanced distribution of database peptide search workload across distributed-memory supercomputers.

Load balance The measure of workload balance between simultaneously running parallel computers working together to compute a solution.

Locality The measure of the closeness of the data to its processing element in the memory hierarchy.

Lock-step A parallel computer architecture where all parallel cores share a program counter thus perform the same operation on a different piece of data all simultaneously and synchronously.

Machine learning An artificial intelligence (AI).

Map-Reduce A computational model consisting of two steps i.e. map and reduce working a divide and conquer computing approach for big data sets often on a parallel computing system.

MassIVE A repository for experimental MS/MS data.

Memory contention A measure of the amount of memory bandwidth under consumption.

Message A data or information exchange between processing elements or CPUs.

Microbiome Microbiome is the collection of all bacteria, fungi, viruses and their genes that live inside our bodies.

Mod Distance (MD) A post-translational mutation aware algorithm to measure similarity between two peptide sequences.

Index A data structure that allows quick querying of information.

Moore's Law Moore's Law states that the number of transistors in a dense integrated circuit (IC) doubles roughly every two years. In today's world, Moore's Law is almost outdated due to thermal limitations of ICs.

Next Generation Sequencing (NGS) Modern and extremely fast DNA sequencing technology which has enabled the production of billions of DNA reads in a few hours.

Open-search The execution of database peptide search for peptide identification using a wide (often ±500Da) precursor mass tolerance setting. Open-search is commonly used to identify spectra with unknown post-translational modifications.

Overfitting A scenario in machine learning where the trained model loses its generality and ends up simply memorizing the data set used to train it. Overfitting is evident if there is a huge gap between training and test set accuracies.

Pipeline stall A scenario in pipeline based parallel computing design when the relative speeds of producer and consumers don't match resulting in a halt in the pipeline flow.

Parallelization The method of re-expressing the algorithmic work such that it could be computed on a parallel computer often faster than a serial computer without compromising on final results.

PSM Peptide to Spectrum Match (PSM) represents a match between an experimental MS/MS spectrum and a peptide or protein sequence. The PSMs are often statistically analyzed for p- and q-value confidence scores.

Post translational modifications (PTM) The mutations that occur in peptides or proteins after they have been synthesized from the genes. PTMs can also be chemically induced during mass spectrometry.

SIMT Single Instruction Multiple Threads (SIMT) is a parallel computing model native to GPUs. In this model, all threads perform the same instruction in a lock-step fashion on a different piece of data.

Non-model proteomics A sub-branch of proteomics that studies the proteomics aspect of the organisms whose reference protein database is not yet available.

Peptide coverage A measure of how well the peptide ion has been dissociated inside mass-spectrometer. Ideally, the peptide ions should break at all possible cleavage points to produce a high quality fingerprint.

Processing Element (PE) A computing element such as CPU, GPU etc.

Pearson coefficient Pearson coefficient is a measure of linear relationship between two sets of data.

Peptide A smaller protein (amino acid chain).

Random sampling A sampling method that selects samples from the original data set at random. The randomization process typically follows a probability distribution.

Regression A statistical method to compute and measure the strength of relation between a given function and a data set. It is also commonly used to model a data set using a known polynomial or a function.

Scalability A measure of parallel computing efficiency depicting the amount of speed advantage that could be achieved by introducing a certain number of processing elements in computing.

Search-engine An algorithm that performs and optimizes a query (experimental spectrum in our case) against a reference database.

Sequencing The process of identifying and assigning a peptide/protein sequence to an experimental spectrum.

Six-frame translation In-silico method for translating a DNA sequence into amino acids taking in account the three possible reading frames in each direction. i.e., 5'UTR and 3'UTR Resulting in six possible translations.

Spectral Library Search Peptide identification method where the experimental MS/MS spectra are compared against a library of pre-identified experimental MS/MS spectra to find the best possible match.

Speedup The measure of increase in processing speed achieved by introducing more processing elements in the system.

SWATH Sequential Window Acquisition of All Theoretical Mass Spectra (SWATH) is a technique for Data Independent Acquisition (DIA) in mass spectrometry.

Index

F. Saeed and M. Haseeb, *High-Performance Algorithms for Mass Spectrometry-Based Omics*, Computational Biology 34,
https://doi.org/10.1007/978-3-031-01960-9